26 Advances in Biochemical Engineering/ Biotechnology

Managing Editor: A. Fiechter

Downstream Processing

With Contributions by
D. J. Bell, P. Dunnill, K. Esser, E. Flaschel,
M. Hoare, J. Holló, M.-R. Kula,
Ch. Lang-Hinrichs, L. Nyeste,
M. Pécs, B. Sevella, Ch. Wandrey

With 72 Figures and 18 Tables

 Springer-Verlag Berlin Heidelberg GmbH 1983

ISBN 978-3-662-38794-8   ISBN 978-3-662-39694-0 (eBook)
DOI 10.1007/978-3-662-39694-0

© by Springer-Verlag Berlin Heidelberg 1983

Originally published by Springer-Verlag Berlin Heidelberg New York in 1983.
Softcover reprint of the hardcover 1st edition 1983

Library of Congress Catalog Card Number 72-152360

2152/3020-543210

For the past twelve years, the series *Advances in Biochemical Engineering* has published high quality scientific information in the areas of both engineering and biology. More recently, these two areas have been successfully integrated into the field known as biotechnology, a subject of growing importance due to the rapid progress being made in applications of molecular genetics and genetic engineering. In order to acknowledge the value of this rapidly-developing technology, the series will now be known as *Advances in Biochemical Engineering/Biotechnology*. In principle, the publishing policy will not change; this change merely reflects our commitment to continue providing the most up-to-date information on recent scientific developments and their practical applications. In the future, too, *Advances in Biochemical Engineering/Biotechnology* will report the latest research progress taking place in the applied fields of biochemistry, molecular biology, microbiology, and engineering, as well as in biotechnology.

Zürich, January 1983   A. Fiechter

Table of Contents

The Formation of Protein Precipitates and Their Centrifugal Recovery

D. J. Bell[1], M. Hoare and P. Dunnill
Department of Chemical and Biochemical Engineering,
University College London, Torrington Place, London WC1E 7JE, UK

---

[1] Present address: Industrial Processing Division, DSIR, Private Bag, Petone, New Zealand

The methods of precipitating proteins by the addition of reagents are outlined from a physico-chemical viewpoint prior to a more detailed consideration of the molecular and colloidal processes by which precipitation occurs. The mixing necessary for efficient contacting of the reagent and protein and for development of the precipitate also generates forces which can break down precipitate particles. The factors which influence the balance between these opposing processes are analysed as one essential element in reactor design. Other elements, including the influence of reactor configuration, are dealt with and the effects of ageing and other precipitate conditioning methods are described. After outlining the options available for precipitate recovery, centrifugal separation is examined in detail and related to the factors which have been shown to influence precipitate characteristics. Several typical industrial precipitation and recovery operations are described and the key factors influencing process design are summarized.

1 Introduction

Protein precipitation and the subsequent recovery of the precipitate from the mother liquor represents one of the most important operations for the laboratory and industrial scale recovery and purification of proteins. These proteins include vegetable and microbial food proteins, human and animal blood plasma proteins and enzymes for analytical and industrial applications. Proteins produced by genetic engineering are also being recovered and purified by this technique.

The term precipitation will be used in this review to describe an operation in which a reagent is added to a protein solution which causes the formation of insoluble particles of protein. In most applications the intention is to recover the protein in either an unchanged molecular form or one which is readily returned to that form. Therefore, reagents such as urea or some metal ions, which induce major irreversible changes in protein structure and can result in precipitation will not be considered. However methods used in the food industry which precipitate and may alter structure will be mentioned.

There are a number of other ways of forming insoluble protein from protein in solution. Heating may be employed either to coagulate the protein or to bring about drying. Conversely the temperature may be reduced to near 0 °C to induce insolubility in some proteins. These methods are industrially important but they do not share the operation of reagent-protein contacting which connects all the methods to be described.

The recovery of heat coagulated protein and especially that of cryoprecipitates does share common features with those considered here.

This review is particularly concerned with proteins whose solubility properties are determined largely by their polypeptide structure. The formation of such structures from the same pool of twenty amino acids gives a degree of conformity though, for example, histones which are characterized by having many basic amino acid side-chains are soluble at low pH but insoluble under alkaline conditions. Proteins with relatively small non-peptide groups such as haem also conform but lipo-, nucleo- and glycoproteins often exhibit distinctive solubility properties. Thus glycoproteins are very soluble in aqueous solution due to the hydration of the carbohydrate moiety, whereas lipoproteins are relatively insoluble due to the hydrophobic nature of the lipid component. Such changes can also be brought about artificially: the reversible binding of the large heterocyclic apolar molecule, rivanol, (2-exthoxy-6:9-diamino acridine lactate) to a protein reduces its aqueous solubility [1] and the reagent has been used commercially for fractional precipitation of plasma proteins.

Proteins were first classified at the end of the 19th century by their solubility properties and though imperfect the terms adopted are still used. Thus the 'albumins' are characterized by solubility in water and in dilute aqueous salt solutions whereas the 'globulins' though soluble in dilute aqueous salt solutions are insoluble (euglobulins) or sparingly soluble (pseudoglobulins) in water.

The structure of proteins is of central importance in appreciating how precipitating reagents function. The polypeptide chain of water soluble proteins will be folded in

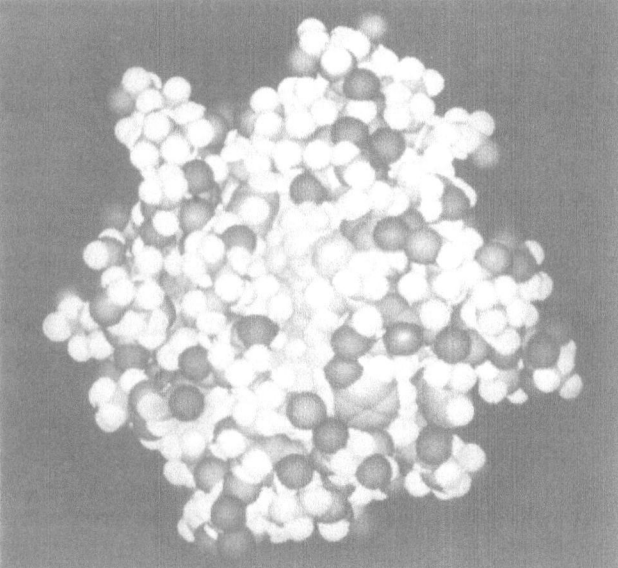

Fig. 1. Model of a globular protein, ferricytochrome c (Tuna). Dark grey spheres are polar atoms, white spheres are hydrogen atoms and pale grey spheres and part spheres are carbon atoms. Thus white and pale grey areas separate from dark grey areas indicate hydrophobic regions. (From computer generated pictures prepared by Dr. R. Feldmann, National Institutes of Health, Bethesda, USA)

such a way that the majority of polar hydrophilic amino acid side-chain groups will be on the exterior and the hydrophobic ones buried (Fig. 1) but this division can never be complete so that changes in the exterior environment brought about by the reagent will affect both types of group as well as the backbone. The overall effect on the protein results from the sum of individual effects which will often be opposed to one another. The situation is complicated by the fact that a protein structure prior to precipitation will not generally be much more stable than a large number of related but undesirable structures. Addition of a reagent can alter the energetic balance in favour of these other structures. Given the complexity of chain folding it will be impractical under industrial conditions to refold the protein to the original structure.

The molecular weights of globular proteins can range from a few thousand to over a million, giving equivalent spherical diameters up to several nanometres. Most globular proteins are ellipsoidal to various degrees. They can often exist as oligomers and this quaternary level of structure may be related to aggregation during precipitation as referred to in this review. Proteins are associated with varying levels of tightly or loosely bound water which forms a hydration shell around the molecule. Up to 0.35 g of water may be tightly bound to 1 g of protein and the amount of loosely bound water is variable but probably stretches up to a few tenths of a nanometre from the molecule surface [2].

A number of biochemical engineering aspects of precipitation are important in establishing efficient operation on an industrial scale. One objective will be to minimize the extent of that part of reagent damage which is due to high localized concentrations. During and after their formation, protein precipitates are subject to particle break-up and further growth according to the hydrodynamic forces to which they are subjected. Such changes may occur during mixing or settling in tanks, passage through pumps and pipelines, and in the equipment used for precipitate recovery. An objective of biochemical engineering studies is to define the causes of these changes in particle size and, where possible, to influence them in a favourable way. During mixing, proteins may be altered irreversibly at gas-liquid interfaces. Their nature is such that foaming readily occurs and this amplifies the possibility of structural damage. Careful equipment design is essential.

The density difference between protein precipitates and the liquid is always small. This, combined with the relatively small particle sizes which can be achieved and the significant enhancement of viscosity which may be caused by proteins and some reagents, leads to problems of recovery. They are made more severe by the hydration and compressibility of the precipitate particles.

Proteins of practical interest are often complex mixtures, many of the components of which have rather similar properties with respect to precipitation. When the objective is to purify a given component protein in the mixture as well as to recover it, precise fractional precipitation is required. The design of a system to achieve this precision is especially demanding.

The biochemical engineering aspects of precipitation will be illustrated by examples from the relatively few systems which have been studied in any detail. They also will be related to engineering design aspects of precipitation reactors, and centrifuges used for precipitate recovery.

Protein precipitation and recovery has some close similarities to the coagulation and flocculation of biological cells and their separation [3]. The precipitation and recovery of most highly hydrated biological substances also show parallels. Defining protein precipitation in biochemical engineering terms therefore should assist in the transfer of information between these related fields.

2 Physico-chemical Aspects of Protein Precipitation

The methods by which proteins are precipitated can be divided into two groups. In the first, protein solubility is reduced by the addition of high concentrations of reagent which change the nature of the solvent environment in a major way. This category includes organic solvents such as ethanol, acetone and ether and neutral salts such as ammonium sulphate. In the second group, which includes acids, bases and some metal ions, low concentrations of reagent are effective by direct interaction with the protein.

The precipitation methods to be described can be summarized as follows:

a) The addition of high concentrations of neutral salts generally decreases protein solubility, an effect known as 'salting-out'.

b) The ionization of the weakly acidic and basic amino side-chains of proteins is influenced strongly by pH. Solubility is a function of the net charge on these groups and proteins have zero net charge at some pH known as the isoelectric point where they will tend to precipitate from solution.

c) If the dielectric constant of the aqueous medium is reduced, for example by the addition of miscible organic solvents, electrostatic interaction between protein molecules is enhanced and precipitation will result.

d) Proteins are precipitated by non-ionic polymers probably as a result of a reduction in the amount of available water for their solvation.

e) A number of charged polyelectrolytes are capable of precipitating proteins, probably by acting as flocculating agents under appropriate pH conditions.

f) Several polyvalent metal ions which interact directly with proteins have proved valuable in reversibily precipitating proteins.

As would be expected, combinations of reagents produce interactive effects. In addition to the above methods which have been widely applied there are a number of reagents such as rivanol which have narrower though sometimes important applications [4].

It is evident from studies with several precipitants that protein solids probably constitute a protein rich phase rather than a pure solid phase such as occurs with, say, sodium chloride crystals and this complicates the analysis of precipitation [2]. A coherent theory describing the action of all protein precipitants is not available. However, the manner in which the major precipitants function will be described briefly since it bears closely on biochemical engineering aspects of precipitation. For more detailed accounts of the physical chemistry of protein precipitation the reader is referred to several reviews [4, 5].

2.1 Colloidal Stability of Protein Suspensions

The colloidal stability of protein suspensions may best be discussed using the Derjaguin London Verwey Overbeek (DLVO) theory. Forces of attraction and repulsion between lyophobic colloidal spheres of uniform charge or uniform potential may be summed and yield an energy barrier to aggregation. The magnitude of this energy barrier relative to thermal energy of the collisions determines whether or not aggregation will proceed and at what rate. This subject has been extensively reviewed elsewhere [6], and only aspects pertinent to protein precipitation will be discussed here. The feature of protein precipitation not commonly considered in the study of lyophobic colloids is the barrier to aggregation due to a hydration shell [7, 8] which will be discussed in Sect. 2.1.2.

2.1.1 Surface Chemistry of Proteins

A protein may be characterized as a globular, amphoteric polymer with a large number of non-uniformly distributed titratable surface groups [9]. The density of surface charges is often low even for a maximally dissociated protein. For example egg albumin, molecular weight of approximately 40,000, has no more than 28 surface charges [10]. The interaction between these surface groups, the proximity of hydrophobic residues or their areas at the surface and buried within the globular molecule and the presence of hydrogen bonds make the prediction of the surface properties of an individual protein molecule a difficult matter [9]. Proteins may exhibit isoelectric points at pH values ranging from 1 to 12 but, for many proteins, for example soya, casein, some albumins, this range is reduced to pH 4 to 6. Therefore in their normal environment these proteins tend to have an overall negatively charged surface. The picture is complicated by the association and binding of various ions and molecules to the protein surface [9, 11].

An overall negative charge of a protein molecule suspended in an electrolyte will attract positive ions from solution to form a layer (Stern) of counterions close to the molecule surface. The more diffuse layer (Gouy-Chapman) of counter-ions then extends further into the solution developing the overall diffuse electrical double layer surrounding the protein molecule. Characterization of the Stern and Gouy-Chapman layers can be related to the zeta potential [12]. This is the potential at the slip plane between the protein molecule and the solution, the plane occurring somewhere in the diffuse layer surrounding the molecule. For a stable non-aggregating system zeta potentials of the particles of the order of ± 10 to ± 40 mV are commonly reported [12]. One widely studied system is a stable suspension of casein micelles where values of zeta potential ranging from -8 mV [13] to -47.6 mV [14] have been reported. The diffuse and irregular nature of the micelle surface makes experimental analysis difficult [15].

The thickness and nature of the Stern layer is dependent on the size and hydration of the adsorbed ions or surface active substances. It is this layer which controls the ultimate approach of the aggregating species. The Stern layer thus indirectly controls the stability of the colloid by control of the size and magnitude of the diffuse layers. At protein concentrations normally encountered in practice, the proximity of neighbouring protein molecules will also exert considerable influence on the structure of the diffuse layers.

The position and structure of the Stern and diffuse layers can be significantly altered by the presence of adsorbed polymers. Generally if the polymers adsorb to the protein surface, they will do so at several points giving a tightly bound species which will not desorb on dilution. The adsorbed polymer will extend a considerable distance from the molecular surface due to the presence of loops and trains, so providing a steric barrier to aggregation and for some polyelectrolytes an increased repulsive force due to electrical interactions. Charged or uncharged polymers affect the Stern layer by reducing its charge intensity, the diffuse layer by rearranging the counter-ions, and the zeta potential by altering the slipping plane between the particle and the solution. The method of addition of the polymers and the concentration added will determine whether or not they act as a colloid stabilizing agent or as a flocculation agent (see Sect. 3.1.3).

Despite the many reservations noted above for characterizing protein molecular surfaces, at least a qualitative picture may be built up for the forces controlling the colloid stability or precipitation of protein molecules. As two similarly charged colloidal species are brought together, the electrical diffuse layers will interact giving rise to a repulsive force between the particles. The outer part of the two layers will very rapidly equilibrate so that the rate of approach may be related to repulsive and attractive forces only. The ions in the electrical layer close to the particle surface, including the Stern layer and the inner part of the diffuse layer, will only relax or equilibrate slowly and effective collisions may be dependent on the orientation of the particles and the nature of the species adsorbed to the surface.

Theoretical expressions have been derived for the idealized case of two uniformly charged colloidal spheres approaching each other. The repulsive force is roughly proportional to the square of the zeta potential of the particle, and decreases exponentially with distance between the particles. Attraction between colloidal particles is due to London-van der Waals forces based on induced dipoles between the colliding particles. These are long range forces operating over several tens of nanometres which is also the case for electrostatic repulsion forces at low concentrations of electrolyte. Ionic bonds are very short range and will only come into play when the repulsive forces have been largely overcome.

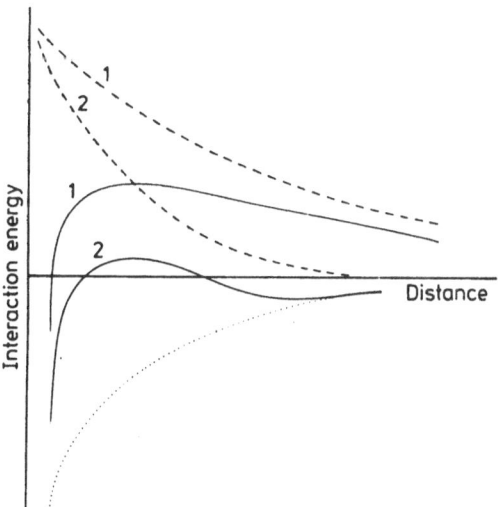

Fig. 2. Combination of attractive and repulsive forces between two charged particles giving overall interaction curves. – – – – electrostatic repulsion forces; · · · · · · London-van der Waals attractive forces; ——— overall interaction energy; (1) low salt concentration; (2) high salt concentration

Considering only the London-van der Waals and the electrostatic forces an overall potential energy barrier to collision may be postulated (Fig. 2). Provided the thermal energy of the particles is sufficient to overcome the overall potential energy barrier then aggregation will take place, the rate of collision being dependent on the relative magnitude of these energy levels. A shallow secondary minimum is noted for the second of the overall energy curves giving rise to loose aggregates which may be easily disrupted on stirring [6].

2.1.2 Hydration of Proteins

Protein molecules cannot be treated simply as hydrophobic colloids. They are very hydrophilic being associated with up to three times their own weight of water. The loosely associated water forms a stabilizing barrier to aggregation [7]. This barrier is not taken into account in the DVLO theory for colloid stability (Sect. 2.1.1).

The loosely held water is not quantifiable due to its similarity to non-associated water. Ten to 100 molecular layers of this water have been estimated to surround a protein molecule probably clustered around the ionic side chains. Some structural ordering would be expected to extend up to a few tenths of a nanometer from the protein molecule surface [2], thereby inhibiting molecules from approaching sufficiently close to promote aggregation.

2.2 Methods of Precipitation of Proteins

2.2.1 Salting-out

The effect of high salt concentration in promoting aggregation and precipitation of proteins is still not well understood. Theoretical approaches have been applied to predict the behaviour but at the high ionic strength involved they cannot be applied strictly and an empirical expression due to Cohn [5] is most widely used:

$$\log S = \beta - KI \tag{1}$$

where S is the solubility of the protein at ionic strength I and β and K are constants. Figure 3 shows an experimental example of protein precipitation obeying the relation and illustrates that if protein concentration is increased the salt concentration at which the equation applies is shifted in the expected manner. The constant, β, of Eq. (1) varies markedly with the protein but is essentially independent of the salt; β is strongly dependent on pH and temperature, usually passing through a minimum at the isoelectric point. The slope of the salting-out curve, K, is found to be independent of pH and temperature but varies with the salt and protein involved. As has been observed with pectins and other hydrophilic polymers, proteins of similar chemical composition but increasing molecular weight require less salt for precipitation. Myoglobin of low molecular weight, albumins, serum globulins, and high molecular weight fibrinogen represent such a series [16, 17]. The more asymmetric the protein molecule the greater is the degree of precipitation at equivalent salt concentrations [7]. For a particular salt, K only varies over approximately a twofold range for different proteins being greatest for large and asymmetric molecules. Salts containing polyvalent anions such as sulphates and phosphates have higher values of K than

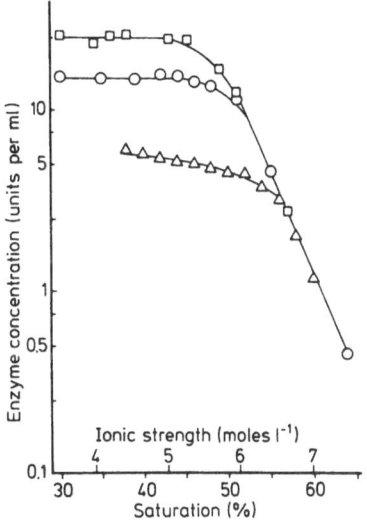

Fig. 3. The effect of enzyme concentration on salting-out of fumarase by ammonium sulphate at 6 °C [112]. Initial enzyme concentrations (units/ml) were: (△) 5.4; (○) 15.2; (□) 25.2

uni-univalent salts but polyvalent cations such as calcium or magnesium depress the value of K. Therefore ammonium and sodium sulphate and potassium and sodium phosphates have been widely used. Sodium sulphate is of low solubility below 40 °C and its use is thus restricted to precipitation of stable proteins such as extracellular enzymes. Although the salting-out effect of phosphates is greater than that of ammonium sulphate the latter is more soluble and its effect is less dependent on temperature, though the variation is still such that temperature must be specified. Ammonium sulphate is also much cheaper. Being the salt of a strong acid and a weak base it does tend to become acid by hydrolysis and the release of ammonia at higher pHs is inconvenient on an industrial scale. Ammonium sulphate is a corrosive material and is difficult to handle and dispose. Residues of it remaining in food products can be tasted at low level, and it is toxic with respect to clinical use so that it must be removed.

The relative effectiveness of neutral salts in salting-out and especially of anions gives rise to a series, the lyotropic or Hofmeister series in which citrate > phosphate > sulphate > acetate ~ chloride > nitrate > thiocyanate. Particularly significant is the observation that the tendency for a salt to cause structural damage to a protein is inversely related to its position in the lyotropic series. Thus sulphate ions are associated with structural stabilization and thiocyanate with destabilization. The specific Hofmeister effects appear to be due to interaction of the ions with hydrophobic groups of the protein [18].

According to a theoretical treatment by Melander and Horvath [19] salting-out of proteins may be described as a balance between a salting-in process due to electrostatic effects of the salt and a salting-out process due to hydrophobic effects. Using a dimensionless form of the Cohn equation (1):

$$\log \frac{S}{S_0} = -KI + \beta' \qquad (2)$$

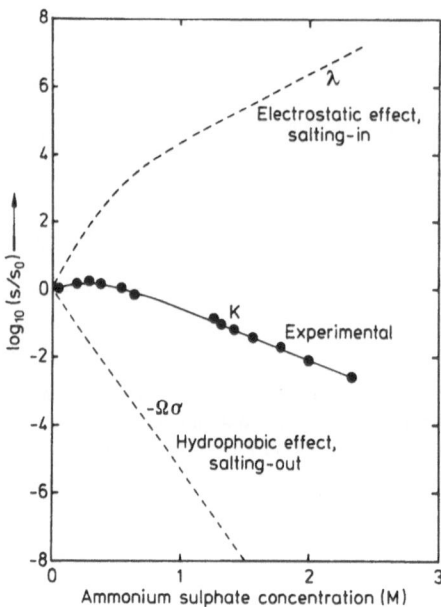

Fig. 4. Contribution of salting-in and the salting-out effects on the overall solubility curve of haemoglobin (see Sect. 2.2.1 for explanation of symbols) [19]

where S_0 is the solubility at zero ionic strength; salting-in, salting-out and overall solubility curves may be summarized as in Fig. 4. The relative surface hydrophobicity determines the contact area between the protein molecules (Ω). At higher salt concentrations the attractive force between the hydrophobic areas is increased due to greater induced dipoles (σ). Concomitant with this increased salting-out effect is the further development of layers of like charges on the molecules, thereby increasing the repulsion between the molecules (λ). The slope of the overall salting-out curve is given by:

$$K = \Omega\sigma - \lambda \qquad (3)$$

The type of salt used determines σ; this property is governed by the molal surface tension increment of the salt. The Hofmeister or lyotropic series of salts reflects their ability to precipitate proteins, and in fact lists the salts in the order of their molal surface tension increment. The protein type essentially determines Ω, and generally the magnitude of $\Omega\sigma$ is much greater than λ. Therefore maximum solubility is reached at relatively low salt concentrations, the solubility decreasing very rapidly at higher salt concentrations. For proteins such as tetrameric myoglobin (molecular weight equals 68,000), ovalbumin (46,000), haemoglobin (68,000) and albumin (67,000), the salting-out curve correlates with the relative surface hydrophobicity or the frequency of charged surface groups rather than with the average hydrophobicity of the molecules. The relationship between the theory of Melander and Horvath and the more common one of high salt concentrations simply removing hydration barriers from around the protein molecule is not clear. Probably the salt does tend to attract water from the protein surface as does the lowering of dielectric constant of the protein environment by the addition of organic solvents, or the addition of hydrophilic polymers such as polyethylene glycol.

2.2.2 Isoelectric Precipitation

Adjustment of pH at low ionic strength to the point where a protein has zero net charge will lead to substantially reduced solubility. Figure 5 illustrates this for soya protein total water extract. The effect, which is enhanced for proteins of low hydration constant or high surface hydrophobicity, has been widely used. For example, casein, a protein with the latter characteristic, precipitates at its isoelectric point forming large, strong, aggregates [7]. On the other hand gelatin, an exceptionally hydrophilic protein does not precipitate at its isoelectric point in low ionic strength or high dielectric constant environments.

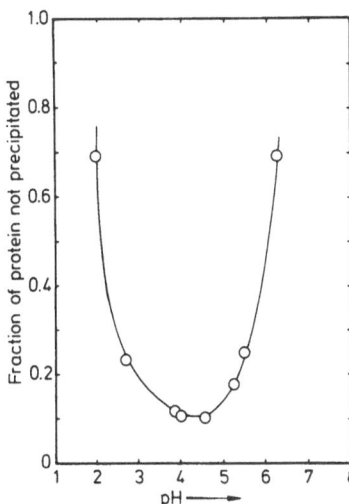

Fig. 5. The effect of pH on the concentration of soya protein remaining in solution, expressed as a fraction of initial concentration of total water extract [99]

The Cohn equation (1) embraces the effect of pH as well as of neutral salts since the constant β for a particular protein is pH dependent, generally reaching a minimum at or near the isoelectric point of the protein. Changes of solubility well in excess of an order of magnitude per pH unit in this region are common. Sometimes β passes through two or more minima, the additional ones evidently being due to formation of specific salts. It is generally observed that as the neutral salt concentration is increased the minimum solubility increases in value and the pH of lowest solubility decreases due to anion binding.

A major advantage of isoelectric precipitation commonly done at acid pHs, is the cheapness of mineral acids and the fact that several such as phosphoric, hydrochloric and sulphuric are acceptable in protein food products. In addition it will usually be possible to proceed directly to another fractionation procedure without the need, faced with salting-out, for removal of the reagent.

The principal disadvantage of using acids is their potential for damaging proteins irreversibly [20]. This comes about because of protein sensitivity to low pH but may be greatly amplified by the action of acid anions from the part of the Hofmeister series (Sect. 2.2.1) associated with protein destabilization.

2.2.3 Reduction of Solvent Dielectric Constant

The addition of a weakly polar solvent to an aqueous solution of a protein will generally result in precipitation and the method has gained considerable importance. For example, with the addition of ethanol whose dielectric constant is about one third that of water, protein solubility decreases considerably due to the increase in effectiveness of intra- and intermolecular electrostatic attractions.

The change in solubility of a protein at its isoelectric point with a change in the effective dielectric constant of the solvent can be described by the expression:

$$\log S = \frac{K}{D_s^2} + \log S_0 \qquad (4)$$

where D_s is the dielectric constant of the reagent-water mixture, K is a constant embracing the dielectric constant of the original aqueous medium and the extrapolated solubility is S_0. Addition of a neutral salt produces a salting-in effect and modifies the value of K [21]. As the solvent dielectric constant is reduced the salting-out effect of the neutral salt decreases more quickly than the salting-in effect and the solubilizing action of small amounts of added neutral salts is used to refine fractional precipitation by ethanol and other solvents. The solubility of metal-protein complexes is very dependent on the dielectric constant of the medium. This property has been exploited in human plasma fractionation where Ba^{2+} and Zn^{2+} salts have been used to give an improved fractionation. It is achieved at lower ethanol concentrations, so reducing protein damage, however destruction of hepatitis virus is less effective.

Precipitation by organic solvents has the advantage that the variable of altered dielectric constant when added to those of pH, temperature, ionic strength and protein concentration gives a very refined method of protein fractionation. Solvents such as ethanol can be obtained by fermentation and distillation as very pure reagents and their volatility permits ready removal and recovery. The bactericidal properties of solvents such as ethanol are also valuable. The principal disadvantage of organic solvents is their tendency to cause structural damage [22]. This demands low temperature operation and, for example, ethanol precipitation of human plasma proteins is undertaken at temperatures as low as -10 °C. The influence of dielectric constant on solubility is very temperature sensitive and the temperature must be held constant (preferably to ± 0.5 °C). Because proteins of higher concentration are precipitated at lower concentrations of solvents of low dielectric constant, there is an advantage in using these conditions to reduce the damaging effects of such solvents. The flammability of the lower alcohols presents a particular problem for large-scale operation where the need for flame-proof plant will raise capital costs. The operating hazards are reduced by employing solvents such as iso-propanol but at the expense of an increased tendency for the proteins to be damaged by the greater hydrophobic character of the solvent. Ethanol suffers the special problem of Government control over its distribution and use. Methanol has been examined as an alternative to ethanol in human plasma fractionation but, with its somewhat higher dielectric constant, precipitation of a given protein is less specific. Diethyl ether was used in England at one time for human plasma fractionation but poor miscibility limits its use to a concentration of 18.5 volume % [23].

2.2.4 Precipitation by Non-ionic Polymers

This method of precipitating proteins has attracted increasing interest recently. Among the polymers most studied have been dextrans and polyethylene glycols. The mechanism of precipitation is not fully explained but the work of Laurent [24] and Ogston and their co-workers [25] suggests that the polymers exclude the proteins from part of the solution and reduce the effective amount of water available for their solvation. The phenomenon is closely related to the formation of liquid-liquid two phase systems from nixtures of aqueous polymers first studied by Albertsson [26] and more recently by Kula and co-workers [27]. Given this close relationship the following expression derived for such systems is expected to apply:

$$\mu_i = \mu_i^0 + RT(\ln m_i + f_{ii}m_i + f_{ij}m_j) \tag{5}$$

where μ_i^0, standard chemical potential of component i; μ_i, chemical potential of component i; R, gas constant; T, absolute temperature; m_i, m_j, molality of component i, j respectively; f_{ii}, self interaction coefficient for component i; f_{ij}, coefficient for the interaction between components i and j.

For polyethyleneglycol precipitation of yeast intracellular proteins Foster et al. [28] showed that precipitation could be expressed by a simplified equation:

$$\ln S + fS = X - aC \tag{6}$$

where $X = (\mu_i - \mu_i^0)/RT$, S, the protein solubility; C, polyethyleneglycol concentration; f, the protein self-interaction coefficient; a, polyethyleneglycol interaction coefficient. The protein self-interaction term was found to be significant at higher protein concentrations $(>10 \text{ mg ml}^{-1})$ and at pH values away from the protein isoelectric point. Protein solubility is markedly influenced by pH, temperature and ionic strength and Miekka and Ingham [29] have shown that some at least of these effects are due to self-association of the protein.

It is observed that high concentrations of polyethylene glycol are required to precipitate proteins of low molecular weight and *vice versa*. Non-ionic polymers of higher molecular weight have greater effectiveness up to a limiting value which for polyethylene glycol is several thousand. These observations are consistent with precipitation by a mechanism of exclusion of the protein from a part of the solvent similar to that occurring in gel filtration. Here, however, it is not necessary to have two visibly distinct phases. Ingham [30] found that albumin solubility with respect to polyethylene glycol was at a minimum at the isoelectric point. He and others have observed only a small influence of temperature with this reagent. Non-ionic polymers have the advantage as precipitants that they stabilize proteins and may be used at ambient temperatures. Precipitation generally occurs at polymer concentrations less than twenty percent and the precipitant can be added as a concentrated aqueous solution. Whereas salts used for salting-out must be removed prior to any subsequent ion-exchange fractionation, non-ionic polymers can be eluted during binding of protein to the ion-exchanger. There has been concern that a residue of the polymers may remain associated with precipitated protein but 4000 molecular weight material is accepted as a safe precipitant for human plasma. The tendency to protein-protein association may reduce fractionation efficiency. It is known that high concentrations of polyethyleneglycol can adversely affect the performance of certain exclusion media such as Sephadex G-100. If necessary the polyethyleneglycol can be forced to become a separate phase by addition of, for example, 0.4 M phosphate when the protein is largely retained in the aqueous phase [31].

2.2.5 Precipitation by Ionic Polyelectrolytes

These reagents appear to act in a fashion similar to flocculating agents with some salting-out and molecular exclusion action and their use seems to be associated with a greater probability of structural change. However they are employed for the recovery of active enzymes and useful food proteins and are therefore worthy of consideration.

A number of ionic polysaccharides have been used for the precipitation of food proteins [32]. Most of the reported studies have been of the use of acidic poly-saccharides such as carboxymethylcellulose, alginate, pectate and carrageenans and they indicate that the major forces responsible for interaction are electrostatic. For example, with carboxymethylcellulose, precipitation occurs below the isoelectric pH of the protein. Imeson et al. provided spectral evidence of diminished thermal stability of myoglobin and bovine serum albumin under conditions of interaction with carboxymethylcellulose [33]. If the protein is deliberately denatured the interaction is increased. Gekko and Noguchi [34] have shown that carboxydextran, carboxymethyl-benzyldextran, dextran sulphate and chondroitin sulphate enhanced the tendency of bovine plasma albumin to thermal denaturation. It is not clear whether structural changes are intrinsic and certainly the use of solid carboxymethylcellulose ion-exchange resins is not associated with damage. As with other flocculating agents an excess of polysaccharide can lead to peptization and resolubilization. A high molecular weight dextran sulphate has been used for the precipitation of serum proteins [35] but has not been applied to fractionation of material for clinical use.

Hill and Zadow [36] have examined the use of the anionic polymers, polyacrylic acid and polymethyacrylic acid and the cationic polymers polyethylene-imine and a polystyrene-based quaternary ammonium salt for the precipitation of whey proteins. Figure 6 shows the patterns of precipitation with respect to pH variation. Polyacrylic acid precipitated more than 90% of the protein from undiluted casein whey at pH 2.8 and the polystyrene-based quaternary ammonium salt precipitated 95% of protein at pH 10.4. Both complexes remained insoluble at ionic strengths of 0.5.

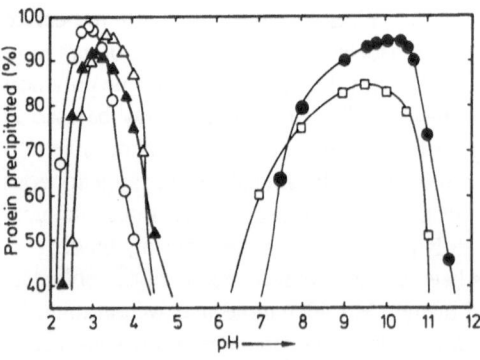

Fig. 6. The influence of pH on the percentage of protein precipitated by various polymers from whey diluted 1:1 with water (ionic strength = 0.1): (●), polystyrene-based quaternary ammonium polymer; (□), polyethyleneimine; (o), polyacrylic acid; (△), carboxymethyl cellulose, degree of substitution 1.5; (▲), polymethacrylic acid [36]

Polyethyleneimines which are widely used as flocculating agents in the clarification of waste water have proved useful in enzyme purification. Vetter et al. [37] found that the unsubstituted product of molecular weight 40,000–60,000 was best and the reagent, a viscous water-soluble product, was introduced into 50 of the 140 different enzyme preparations done at the Boehringer Mannheim company and patents were filed [38]. Polyethyleneimines complex with anionic substances and precipitation is strongly affected by the type and concentration of any salt anions present as well as by pH. The enzyme can be recovered by extraction with strong salt solutions and polyethyleneimine removed by complexing with the salt anions or by cation exchange.

Sternberg and Herschberger [39] used polyacrylic acid to precipitate lysozyme. Separation of the component of the complex was achieved by precipitation of the polyacrylic acid as its calcium salt at pH 6. The recovery of lysozyme activity was 92.3 %. Tannic acid is used industrially for the precipitation of extracellular enzymes where it is associated with very considerable increases in enzyme concentration. Greenwalt et al. [40] used it to purify urinary proteins 500-fold. Polyphosphates are known to precipitate plasma proteins other than γ-globulins at a pH between 4 and 5. For example, a reagent $(H_3PO_4)_x$ with about 70 % of the molecules with x greater than 10 has been used and subsequently removed with an anion resin [23]. More recently polyphosphates have been used to precipitate casein and gelatin [41]. Sternberg examined the use of the heteropolyacids; tungstophosphoric, tungstosilicic and molybdosilicic acids, as protein precipitants [42]. Precipitation occurred just below the isoelectric point and was evidently brought about by the interaction of the heteropolyacid anion with increasingly electropositive protein. Cerbulis has examined the use of bentonite, a cation exchange clay mineral, and lignosulphonate as precipitants for cheese whey and found them more effective than hexametaphosphate [43].

2.2.6 Precipitation by Metal Ions

Specific effects associated with particular anions have been discussed with respect to salting-out. The distinction between precipitating effects of different metal ions is even more pronounced. A number of polyvalent metals are effective in precipitating proteins [4, 44]. They may be classified into three groups. Ions such as Mn^{2+}, Fe^{2+}, Co^{2+}, Ni^{2+}, Cu^{2+}, Zn^{2+} and Cd^{2+} bind strongly to carboxylic acids and to nitrogenous compounds such as amines and heterocyclics. On the other hand Ca^{2+}, Ba^{2+}, Mg^{2+} and Pb^{2+} bind to carboxylic acids but not significantly to nitrogenous ligands. A third group, including Ag^+, Hg^{2+} and Pb^{2+}, bind strongly to sulphydryl groups. The ions Ca^{2+}, Ba^{2+} and Zn^{2+} have been particularly useful in modulating ethanol precipitation of human plasma proteins. An advantage of some metal ions is their greater precipitating power with respect to proteins in dilute solution. Metal salts employed in low concentrations for separating proteins may be removed subsequently by ion exchange resins or by chelating agents.

2.3 Fractional Precipitation

As well as its potential for concentration, precipitation has proved valuable for fractionation of proteins from one another. This use is much more demanding and it is a common experience that published results are hard to reproduce and that resolution is reduced on scale-up.

The theory of fractionation by salting-out has been described in detail by Dixon and Webb [45] and is dealt with in other useful reviews [4, 5, 46]. Holding protein concentration constant is important if reproducible results are to be achieved. For a protein concentration change from C_1 to C_2 the ionic strength at which precipitation occurs will change from I_1 to I_2 where:

$$\log (C_1/C_2) = K_s(I_2 - I_1) \tag{7}$$

Dixon and Webb provide some evidence of lack of interaction in the precipitation of proteins at high salt concentrations [45]. They also note the virtues of operating at higher protein concentrations: these stem from precipitation at lower salt levels which in turn reduces cost and eases centrifugal recovery of the precipitate by avoiding excessive specific gravity increase of the liquor. The optimum range of salt concentration to recover an individual component protein from a mixture is concluded to be 5–10% of saturation for ammonium sulphate. Dixon and Webb recommend precipitation of 75% of a given protein to obtain a good yield consistent with high purity but the proportions will be varied according to the importance of full recovery and the availability of other subsequent purification stages.

In isoelectric precipitation, as noted, the solubility of an individual protein near the isoelectric point may change by over an order of magnitude within a single pH unit. Given a difference of isoelectric points for two proteins of two units which is not uncommon, the ratio of their solubilities may be changed several thousandfold over one unit of pH. However the independence of precipitation experienced with salting-out even at relatively high protein concentrations is not so evident for isoelectric precipitation.

Large-scale fractionation by manipulation of dielectric constant provides an impressive practical example of fractional precipitation, namely for the ethanol fractionation of human plasma proteins, (Fig. 7). The need for it stems from the scarcity of the human plasma feedstock and the consequent need to make best use

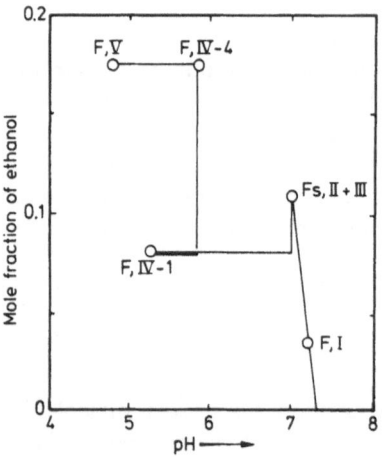

Fig. 7. The variation of ethanol concentration and pH during the fractional precipitation of human plasma by Method 6 due to Cohn et al. [47]. Therapeutic fractions, F, are indicated

of each of the key protein components. Because the supernatant from one fractional precipitation is the starting material for the next, aberrations at any point are carried forward: these may include under- or overprecipitation and structural damage. Though variation of ethanol concentration is a central element, use is made of pH, temperature, ionic strength and protein concentration changes [47].

Fractionation efficiency with polyethylene glycol was found to be impaired by protein-protein interaction at higher protein concentrations where the protein-protein term of Eq. (6) becomes significant [28]. Honig and Kula [48] have noted that nominal 300 molecular weight polyethylene glycol was superior to that of higher molecular weight (4000–6000) with respect to selectivity of precipitation.

Carboxymethylcellulose has been used to fractionally precipitate β-lactoglobulin and α-lactalbumin [32] and the same method has been used to fractionate yeast proteins [32]. Alameri [49] separated γ-globulin from albumin at pH 5.6, where albumin is negatively and γ-globulin positively charged, by selective precipitation with sodium methacrylate. The precipitate containing the γ-globulin could then be redissolved by raising the pH and the methacrylate removed by precipitation with barium salts. Fractionation using polyethyleneimine can be highly selective (Fig. 8) with a careful choice of pH even when the isoelectric points of enzymes, glucose oxidase (pH 4.3) and catalase (pH 4.6) are close [37].

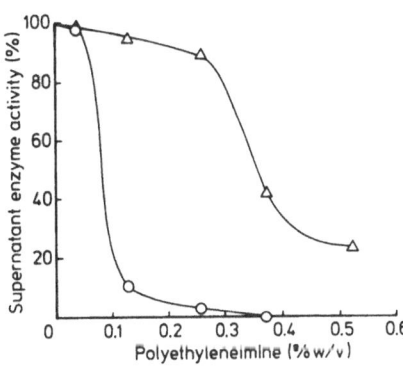

Fig. 8. Fractional precipitation of glucose oxidase (o) by polyethyleneimine at pH 7.6 in the presence of catalase (Δ) [37]

The resolving power of fractional precipitation can often be improved by lowering the total protein concentration which reduces co-precipitation and occlusion. However dilute proteins are more susceptible to unfolding, sub-unit dissociation and oxidation damage.

2.4 Choice of Precipitant

The factors affecting the choice of precipitant can be summarized under several headings.

a) Protein structural stability. Precipitants such as polyethylene glycol and ammonium sulphate stabilize a wide variety of globular proteins whereas others, such as alcohols and some ionic polyelectrolytes and metal salts, are associated with denaturation and must be employed under very precisely controlled conditions which can be more difficult to achieve in industrial environments.

b) Operational ease of use. In addition to restrictions with reagents prone to destabilize protein structure, some reagents pose other operational difficulties. Alcohols and ethers are flammable and ammonium sulphate under alkaline conditions releases ammonia and it is not easy to meter the solid salt reliably. On the other hand polyelectrolytes can conveniently be used as concentrated aqueous solutions. Such operational factors are of great importance in large scale operation.

c) Fractionation. Salting-out in which protein-protein interaction is small is associated with a high degree of resolution whereas isoelectric and neutral poly-electrolytes seem to induce interaction which can have the effect of bringing down protein below the reagent concentrations where this would be expected. The effect is more severe at the higher protein concentrations which are desirable in industrial operation. Since precipitation is often one of the earliest of several fractionation procedures in obtaining purified proteins the influence of residual precipitation reagent on subsequent stages will be important. Whereas removal of salts will often require laborious dialysis or diafiltration prior to chromatography this may not be required for non-ionic polymers. Ionic polyelectrolytes and metal ions can often be removed by secondary precipitation steps carried out on the redissolved protein solution.

d) Product end use. Proteins prepared on a large scale for commercial use must meet more stringent requirements with respect to residues of reagents than are necessarily demanded during use in fractionation. Human plasma protein fractions may be injected and so very careful evaluation of the effect of residues will be essential. Ethanol is attractive since its initial quality is high and it can be readily removed. Non-ionic polymers, metal ions and salts have been permitted but their removal presents greater problems. Ammonium sulphate is readily tasted in food products and in this respect isoelectric precipitation with mineral acids is attractive.

The precipitating action of different categories of reagents has been discussed separately but in practice it is common to use combinations of reagent such as acids and salts, alcohols and metal ions, and polyelectrolytes and acids. Such combinations can enhance performance particularly with respect to fractionation but equally they may lead to the adoption of complicated recipes based on little more than scientific folklore. Though a precise physico-chemical description of the action of protein precipitants is not yet possible their behaviour is sufficiently well understood to be able to subject any proposed procedure to the test of whether it conforms to the expected pattern.

3 Formation of Protein Precipitates

As discussed in the previous section the solubility of protein molecules may be changed by alteration in the pH, ionic strength or dielectric constant of the protein environment or by addition of various metal ions or polymers. Under specific conditions molecular association leading to precipitation of the protein will occur. The kinetics of precipitation is discussed on the basis of the factors controlling the rates of molecular association and precipitate aggregation.

3.1 Protein Association and Aggregation of Precipitates

In a protein solution the random motion or thermal energy of the molecules will promote collisions. Removal of any hydration or electrical barriers to collision will allow association of the protein molecules. The particles so formed will continue to grow by diffusion until they are of a size where fluid motion becomes important in promoting collisions. These two growth processes are termed perikinetic and orthokinetic aggregation respectively.

3.1.1 Perikinetic Growth

In a mono-sized dispersion the initial rate of decrease of particle number concentration (N) can be described according to the theory of Smoluchowski [50] as a second order process:

$$-\frac{dN}{dt} = K_A N^2 \tag{8}$$

where the rate constant, K_A, will be determined by the diffusivity, D, and diameter, d, of the particles:

$$K_A = 8\pi\, Dd \tag{9}$$

These equations apply up to a limiting particle size defined by the fluid motion, this size typically ranging from 0.1 to 10 µm for high and low shear fields respectively [51]. Smoluchowski's theory has been experimentally verified for materials such as suspensions of latex particles [52] and colloidal metal dispersions [53].

Any electrical barrier around the particles will reduce the rate of association and this is accounted for in a modification, due to Fuchs[54], of Smoluchowski's theory:

$$-\frac{dN}{dt} = \frac{K_A}{W} N^2 \tag{10}$$

where W is a stability ratio:

$$W = d \int_0^\infty \frac{\exp\{\varphi(h)/kT\}}{(h+d)^2}\, dh \tag{11}$$

h is the particle separation distance and k is Boltzmann's constant. The potential energy of interaction, $\varphi(h)$, between the two particles may be computed for particles having constant surface charge or constant surface potential. Such interaction terms have been introduced into the perikinetic aggregation theory to allow for interparticle attractive and repulsive forces [8]. Corrections are also needed for hydrodynamic removal of free water between the particles [55] and particle non-sphericity [56]. A simpler, empirical approach has been to incorporate an effectiveness factor to allow for less than the expected numbers of collisions leading to aggregation [51].

Removal of the stabilizing layer around protein molecules should cause protein aggregation at a rate predicted by Eqs. (8) and (9). Any remaining surface charge on the protein molecules will reduce the aggregation rate as in Eqs. (10) and (11). However, neither surface charge nor surface potential will remain constant as the molecules approach due to altered degrees of dissociation of the surface acidic and basic groups. Hence the stability ratio, W, cannot be evaluated readily. The theories described by Eqs. (8)–(11) do not account for the presence of a hydration barrier of associated water around the molecules. Protein association may not be the limiting step in the process of precipitation and this may apply especially at the start of the precipitation process where the removal of the hydration or electric barriers can be very slow processes. Smoluchowski growth kinetics may be related to change in molecular weight, $\bar{M}_w$, of the aggregate with time [57, 58]. A linear relationship is obtained:

$$\bar{M}_w(t) = \bar{M}_w(o) (1 + K_A m_0 t) \tag{12}$$

where m_0 is the molar concentration of the aggregating species. Such a relationship has been observed for the calcium ion induced aggregation of casein (Fig. 9) [59, 60].

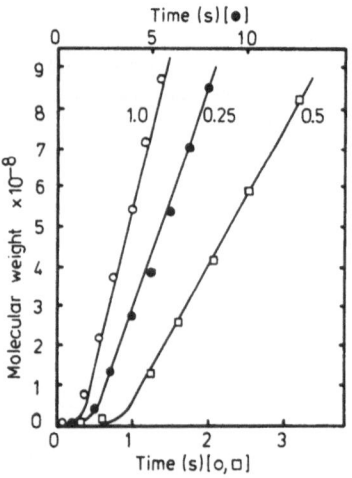

Fig. 9. Molecular weight — time plots for three concentrations of α_s-casein (concentrations as indicated on graph in kg m^{-3}), aggregating in the presence of 0.008 M CaCl$_2$. Lines represent theoretical prediction of growth rate based on modified version of Smoluchowski's theory of perikinetic aggregation [59]

Parker and Dalgleish [59] proposed that in the early stages of the growth process, calcium ions bind to the casein molecules and provide, through the second and free valence on the calcium, multifunctional points for aggregation. The rate of collisions was limited by the number of free valences until aggregation had proceeded to such an extent that precipitate growth was diffusion controlled. No allowance was assumed for any residual repulsion or hydration barrier to collisions. That is, all the collisions were taken to be totally effective. This relationship was followed up to the formation of casein particles of mean molecular weight of 10^9 or particle size of 0.1 μm on the basis of the following relationship between particle diameter, d μm, and mean molecular weight, $\bar{M}_w$ [61]:

$$\bar{M}_w = 0.105 \times 10^{12} d^3 \tag{13}$$

The corresponding relationship for a complex of globular protein molecules has been put forward [62]:

$$\bar{M}_w = 0.235 \times 10^{12} d^3 \tag{14}$$

The enzyme induced coagulation of casein micelles has also been described using Smoluchowski kinetics [13, 63, 64]. In this case the initial growth of the micelles was modelled using Michaelis Menten kinetics to describe the proteolysis of x-casein, a necessary precursor to the association process. Alternatively in the case of various chemically modified caseins a 'coagulation time' may be defined by extrapolation of the linear perikinetic aggregation growth curve of a precipitate back to the time axis, that is the hypothetical point of zero aggregation. The coagulation time (CT) in the presence of calcium ions was shown to be as follows [65]:

$$CT \propto e^{-q^2} \tag{15}$$

indicating that the total energy of interaction between the molecules on the basis of an Arrhenius relationship is a direct function of the square of the net surface charge on the casein molecule, q, a result in agreement with the DLVO theory for the stability of colloids (Sect. 2.1.1).

It is unsatisfactory that the most detailed kinetic studies are of casein aggregation since casein molecules are not typical of globular proteins. Due to the greater colloidal stability of other protein molecules, aggregation may well be slower than for casein.

Smoluchowski's theory has been shown to describe the isoelectric point precipitation (pH 4.6) of urea denatured ovalbumin [66]. Using a Coulter counter to monitor the rate of loss of particle numbers with time, the rate constant for a second order process may be assessed from a plot of the inverse of the particle number concentration versus time. Assuming the Stokes-Einstein relationship for translational diffusion, D:

$$D = \frac{kT}{3\pi \mu d} \tag{16}$$

and fully effective, irreversible collisions, the rate constant for perikinetic aggregation, see Eqs. (8) and (9), of 0.8 to 3 μm diameter particles was 5.6×10^{-18} m³ s⁻¹ as compared with 6.2×10^{-18} m³ s⁻¹ measured experimentally. Raising the pH of the solution to 5.5 was sufficient to give negligible aggregation of the protein molecules.

The growth of precipitate particles of 3 to 12 μm diameter has been studied under conditions likely to favour perikinetic aggregation. Casein precipitate, prepared by continuous salting-out at 1.8 M ammonium sulphate, was aged during hindered settling. The change in particle size distribution of the precipitate phase showed reasonable agreement with theoretical prediction for the first period of ageing where minimal settling occurred, while for larger ageing times increases in viscosity of the thickened sludge probably accounted for considerable deviation from theory [67, 68] (Fig. 10).

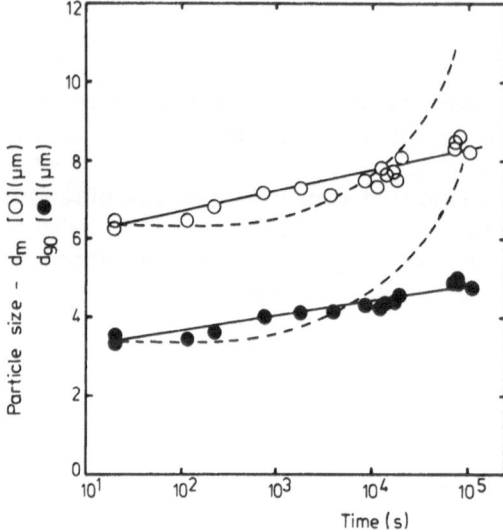

Fig. 10. Particle size change during hindered settling of casein precipitate prepared by salting-out with 1.8 M ammonium sulphate; – – – – theoretical prediction of precipitate growth due to Smoluchowski's theory of perikinetic aggregation [67, 68]

3.1.2 Orthokinetic Aggregation

In a sheared suspension of particles greater than approximately 1 μm diameter, fluid motion will cause particles to collide and hence aggregate. These collisions also can promote aggregate break-up as may collisions with solid surfaces in the particle environment or the action of fluid shear.

For a suspension of uniformly sized spherical particles, diameter d, in a uniform shear field, shear rate G, the initial rate of decrease of particle number concentration, N, due to collisions is given by [50]:

$$-\frac{dN}{dt} = \frac{2}{3}\alpha Gd^3N^2 \tag{17}$$

where α is the collision effectiveness factor or the number of collisions which result in permanent aggregates. Assuming a constant volume fraction of particles, φ_v, during aggregation ($\varphi_v = \pi d^3N/6$) the Eq. (17) may be rewritten:

$$-\frac{dN}{dt} = \frac{4}{\pi}\alpha\varphi_v GN \tag{18}$$

or in terms of the ratio of final to initial particle number concentrations (N_t/N_0) after exposure to shear for time t:

$$\frac{N_t}{N_0} = e^{(-4\alpha\varphi_v Gt/\pi)} \qquad (19)$$

Ideally Eqs. (17) to (19) should also include a term for perikinetic aggregation. However this term will become less important with increasing particle size [51].

Collision effectiveness factors, allowing for such effects as interparticle repulsion and orientation requirements, may be estimated by studying the rate of decrease in particle numbers during perikinetic aggregation. Using such a technique orthokinetic aggregation theory has been shown to predict the growth rate in Couette flow of latex particles [69]. Generally, low collision effectiveness factors are needed to explain the observed growth rates of orthokinetic aggregating systems even in the absence of repulsive and steric barriers to aggregation [55, 70]. These deviations are due to several shortcomings in Smoluchowski's theory for orthokinetic aggregation. Two particles apparently on a direct collision course will tend to follow the fluid stream lines around each other thereby diminishing the chance of collision. Hence the effectiveness of a collision also becomes a function of the shear rate and the particle sizes. In a rigorous analysis of the hydrodynamics of colliding particles, incorporating terms for van der Waals interactions, van de Ven and Mason [71] showed the effectiveness of collisions to be proportional to $G^{-0.18}$ and approximately proportional to $d^{-0.73}$ in the absence of interparticle repulsive forces. Hence the effectiveness of collision may be reduced from 100 to 28% for 2 μm diameter particles exposed to a rate of shear of 50 s^{-1} and from 100 to 6% for 20 μm diameter particles under the same rate of shear [72]. These calculations were made assuming no interparticle repulsion.

For most laminar flow systems there is a variation in shear rate with position in the shear field. For example, in laminar pipe flow, particles near the pipe wall will be exposed to high shear rates for long periods of time, both these factors decreasing further away from the wall. Incorporating varying Gt values and corrected collision effectiveness factors into Smoluchowski's basic theory for orthokinetic aggregation gives reasonable predictions for the growth of polystyrene particles suspended in water [72]. Design on the basis of average shear rates, a common practice in the design of flocculation systems, may cause significant errors. Disagreement between theory and practice also exists due to an inability to correct for the non-sphericity of the resultant aggregates.

Turbulent flow provides additional complications since the shear field is non-uniform and variable at any point with time. A turbulent fluid may be characterized by the turbulence microscale (η) which gives a measure of the size of eddies in the fluid:

$$\eta = \left(\frac{v^3}{\varepsilon}\right)^{1/4} \qquad (20)$$

where v is the kinematic viscosity and ε is the energy dissipation per unit mass of fluid. At least in the early stages of protein precipitation and probably for most complete protein precipitation processes the size of the protein aggregate is smaller than the microscale length. Particle collision rates for this case have been rigorously analysed by Saffman and Turner [73] for neutrally buoyant particles in a homogeneous turbulent flow field. For two particle sizes (diameter d_1 and d_2) the collision frequency is given by:

$$-\frac{dN}{dt} = \frac{1.294}{8} (d_1 + d_2)^3 \left(\frac{\varepsilon}{v}\right)^{1/2} N_1 N_2 \tag{21}$$

which for a monosized distribution reduces to:

$$-\frac{dN}{dt} = 0.647 \, d^3 \left(\frac{\varepsilon}{v}\right)^{1/2} N^2 \tag{22}$$

For homogenous or non-homogenous turbulent flow fields Camp and Stein [74] earlier proposed that the average shear rate in Smoluchowski's equation for ortho-kinetic aggregation, see Eq. (17), should be estimated on the basis of the power dissipation per unit volume:

$$-\bar{G} = \left(\frac{\varepsilon}{v}\right)^{1/2} \tag{23}$$

giving:

$$-\frac{dN}{dt} = \frac{2}{3} d^3 \left(\frac{\varepsilon}{v}\right)^{1/2} N^2 \tag{24}$$

a similar result to that given in Eq. (22). The Camp and Stein approximation, see Eqs. (23) and (24), is commonly used in the design of water treatment flocculation plants and is the basis for the dimensionless Camp number (Ca):

$$Ca = \bar{G}t \tag{25}$$

Equations (21), (22) and (24) should include the normal collision effectiveness factor as described previously and all the reservations in the application of Smoluchowski's theory to aggregation in a laminar shear field equally apply to turbulent shear fields. Also in a non-homogeneous turbulent shear field a spectrum of turbulent microscales will exist. For example in a turbine stirred vessel (height = diameter = 0.15 m, turbine speed ~4 rps) containing water, the average microscale of turbulence in the impeller regions will be approximately 24 µm while the average for the bulk of the fluid will be approximately 56 µm [75]. Argaman and Kaufman [76] have developed a diffusion coefficient for particles or aggregates based on the turbulence energy spectrum. Even in regions of high energy dissipation much of the protein precipitation process will involve particles or aggregates smaller than the average turbulence microscale.

3.1.3 Flocculation [77]

As noted in previous Sects. 2.2.4 and 2.2.5 proteins may be precipitated by the use of non-ionic or ionic polymers. The mechanism of precipitation is uncertain but it is probable that flocculation plays a considerable part especially in the case of ionic polymers.

An understanding of flocculation mechanisms, particularly as applied in the water treatment area, should aid the choice of protein flocculants and the design of systems using polymers. By analogy with cell flocculation [3] two mechanisms may be used to describe the flocculation process. One involves charge neutralization of the protein surface when the flocculating agent and protein species are of opposite charge. This mechanism should be similar to isoelectric point precipitation. The other mechanism involves bridging between particles by the flocculating agent, without necessarily removing or neutralizing the repulsive charge barrier. The polymer will usually be of larger size than the protein to be treated and this mechanism leads to loose flocculated structures.

The rate of flocculation will depend on the mixing of the flocculating agent with the protein solution and the rate of diffusion and the attachment of the flocculating agent to the protein molecule surface. There may also be a need for the flocculating agent to rearrange at the protein molecule surface. All these stages will determine the lag time before flocculation commences. The flocculating agent may continue to attach to further points on the same protein molecule surface or bind to other protein molecules. The relative rates of these reactions will largely determine the degree of bridging and the density of the final flocculated structure.

The rate of collision of polymer colloid species has been described on the basis of a modified Smoluchowski's theory for orthokinetic aggregation [76, 77] (Sect. 3.1.2). Comparison between theoretical and experimental rates of flocculation is often confused by shear break-up of the flocs and the possibility of polymer scission (Sect. 3.3). Also the size of the aggregating species may be significantly increased by the adsorbed polymer [78].

The best flocculating agents often have weakly or strongly ionizable groups, though some are non-ionic. Collagen and gelatine are among the earliest flocculating agents used [79] especially in the food and beverages field, for example, in the removal of proteinaceous beer hazes. For those polyelectrolytes which have weakly ionizable groups, charge density, size and shape and hence flocculating ability, are all strongly affected by their environment.

3.2 Shear Break-up of Protein Precipitates

Break-up of flocs or aggregates may proceed by several mechanisms. These include, bulgy deformation due to pressure gradients across the aggregate, fragmentation, primary particle erosion due to hydrodynamic shear, and particle-particle or particle-surface collisions [80, 81]. The two extremes in characterizing particles with respect to break-up are liquid droplets and brittle solids.

Liquid droplets in a turbulent or laminar flow field deform as a function of the relative velocity across the droplet [82]. For break-up to occur the forces of distortion have to overcome the increased surface energy of the droplet. The droplet will break into two roughly equal parts, possibly also forming very small droplets. Droplet break-up is usually characterized by the maximum stable droplet diameter to exist in a particular shear field [83, 84, 85]. The other extreme in break-up is characterized by the abrasion in liquid suspensions of brittle substances such as crystals [86, 87] or coal particles [88]. Removal of protuberances, corners and edges from the particles proceeds as a first order process with respect to particle concentration. Power dissipated into the suspension causes abrasion by particle-particle and particle-surface collisions, the degree being dependent on the particle hardness.

The properties of protein precipitates place them somewhere in the range between brittle particle abrasion and liquid droplet break-up. Isoelectric point soya precipitate aggregates exposed to a laminar Couette shear rate of 2000 s^{-1} show an initially rapid decrease in aggregate size followed by a slow steady decrease [89]. Continued exposure to shear for up to 50 h does not show attainment of an'equilibrium value expected from a balance between orthokinetic aggregation (Sect. 3.1.2) and shear break-up, indicating the partly irreversible nature of protein precipitate break-up. A similar phenomenom is noted for flocculated particles where the shear break-up of flocs may be accompanied by scission of bridging polymers and increased occupancy of available binding sites and leads to lower effectiveness of subsequent collisions in forming new flocs [90]. Rapid initial break-up of protein precipitate is noted when it is exposed to shear in various types of pumps (Fig. 11) [89]. However, as is emphasized by the peristaltic pump data, it is possible in some cases to design transport systems which do not cause significant shear damage of the protein precipitates.

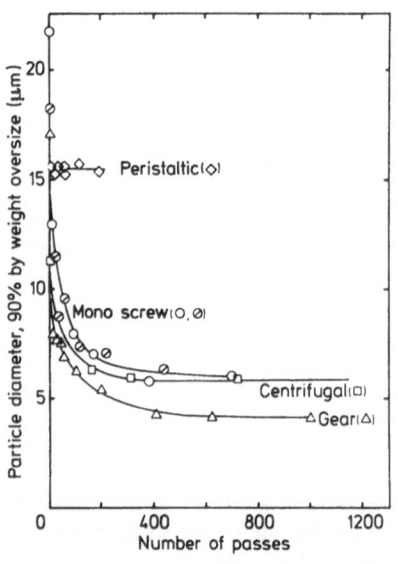

Fig. 11. Particle size change of isoelectric soya protein precipitate after flow through various pumps under total recycle. Total protein concentration = 2.5% by weight [89]

The effect of protein precipitate concentration on the ratio of the particle diameters after and before shear break-up is shown in Fig. 12. For dilute suspensions of protein precipitate, shear controlled growth becomes negligible. The break-up process may also be concentration dependent so that lower concentrations diminish particle-particle and particle-surface collisions [91]. Precipitate exposed to very high shear rates in laminar capillary flow shows this concentration dependence even for shear times of fractions of a second [92].

Examination of the change in protein precipitate particle size distribution during laminar shear indicates that break-up is by fragmentation, the fragment size decreasing rapidly over the first period of shear (Fig. 13) [89]. It is probable that this process proceeds until almost spherical aggregates are obtained, further break-up being due to the erosion of primary 1 μm diameter particles. The fragment size is a very weak function of the rate of shear [92] whereas the rate of breakdown especially of large aggregates is strongly dependent on the shear rate. The effect of these trends on the subsequent precipitate particle size distribution is shown in Fig. 14. The diameter at the large end of the size distribution (d_{05}) shows a strong dependence on the applied rate of shear while the diameter at the fine end (d_{90}) is almost independent of it. The d_{90} value is determined by a combination of the rate of fragmentation and

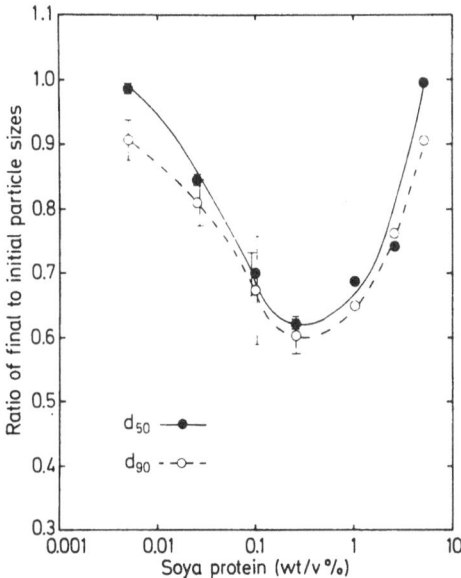

Fig. 12. Effect of isoelectric soya protein precipitate concentration on the fractional decrease in particle diameter after 4 h exposure to a rate of shear of 2000 s^{-1} in Couette flow [91]

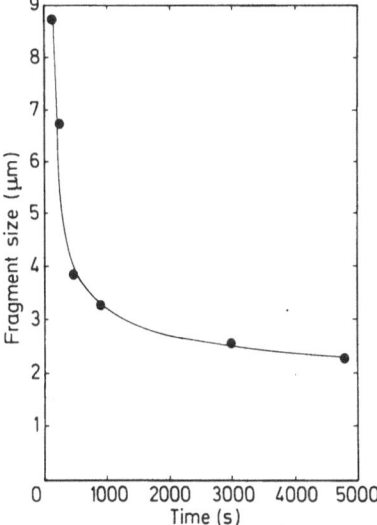

Fig. 13. Effect of ageing time at 2000 s^{-1} in Couette flow on mean fragment size during break-up of isoelectric soya protein precipitate [89]. Initial mean particle size (time, 0 s), 34 μm; final mean particle size (time, 7200 s), 10 μm. Protein concentration, 2.5% by weight; temperature, 23 °C

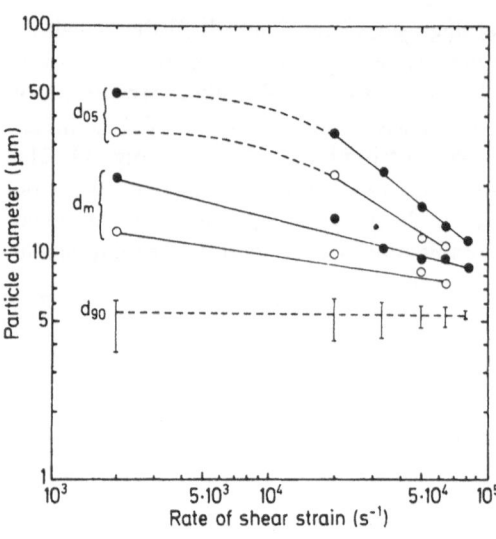

Fig. 14. Effect of rate of shear on isoelectric soya protein precipitate particle size [89]. Time of shearing, 300 s (●); 1800 s (o). Ranges of d_{90} values for times of shearing from 300 s to 7200 s are shown. Total protein concentration in suspension, 2.5 % weight; temperature, 23 °C

the rate of orthokinetic aggregation during which large precipitate particles mop up the fines.

A most important factor determining protein precipitate aggregate strength is the history of preparation. Unless the precipitate is exposed to sufficient deformation forces it will not develop a compact structure. The strength and density of a floc have been shown to be directly related quantities [93]. Optimum strength of soya protein precipitate aggregates is achieved when the product of shear rate and ageing time exceeds 10^5 (Fig. 15). See Sects. 4 and 5 for further discussion of the precipitate properties.

Theoretical or even empirical models for the prediction of shear break-up of protein precipitates have not proved successful due to factors such as the time-dependent properties of the precipitate during shear break-up. Many theoretical or semi-empirical models have been presented for the shear break-up of relatively large

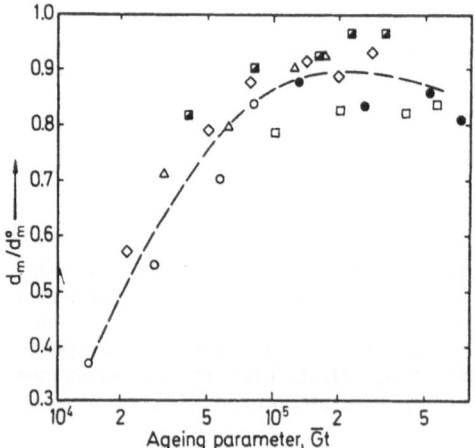

Fig. 15. The ratio of the mean particle size after and before exposure to capillary shear as a function of $\bar{G}t$. $\bar{G}$, reactor mean velocity gradient during precipitate ageing; t, ageing time in reactor [92]. Average capillary rate of shear: 1.7×10^4 s^{-1}; average time of exposure in capillary: 0.065 s; protein concentration: 30 kg m^{-3}; initial mean diameters (prior to exposure to capillary shear), micrometres: o , 53,5; △ , 23.4; ◇ , 19.5; ▣ , 15.2; ▢ , 10.2; ● , 8.8.

flocs of inorganic materials. Earlier descriptions of floc break-up assumed a behaviour approaching that of a liquid droplet in a shear field, where the floc is subject to fluctuating pressure causing bulgy deformation and ultimately disruption [94]. The force required for disruption is a strong function of the plastic or yield stress of the floc suspension [95]. However an alternative mechanism is often proposed to allow for the filamentous structure of the floc, where turbulent drag on the floc causes surface erosion [76]. Complex theories for the various mechanisms, particle dimensions and scales of turbulence have been presented by Parker et al. [81]. They require specific modification and correction for each system [55, 96, 97].

The simplest correlation for break-up usually relates to the maximum stable size of an aggregate which exists in a shear field. Correlations for inorganic aggregates and flocs have been reviewed recently by Tomi and Bagster [75]. Aggregates larger than the turbulent microscale η, see Eq. (20), and hence exposed to break-up by inertial forces due to particle-particle and particle-surface collisions have maximum stable particle diameters strongly dependent on the impeller speed (N_s), mean velocity gradient $(\bar{G})$ or power dissipation (ε) in a stirred vessel. That is, j approaches a value of one in the following equation:

$$d_{max} \propto N_s^{-3j} \propto \varepsilon^{-j} \propto \bar{G}^{(-2j)} \tag{26}$$

For break-up by viscous forces, where aggregates are smaller than the scale of turbulence, j approaches a value of zero. Maximum surface shear effects occur when the particle is about the same size as the turbulent microscale and for this case j is often assumed to be 0.5. Experimental values of j ranging from 0.1 to 0.8 have been noted for flocculation and precipitation processes [93, 98].

3.3 Modelling the Growth of Protein Precipitates

To deal with the wide protein precipitate particle size distributions for modelling purposes more comprehensive forms of Eqs. (8) and (17) are needed to describe perikinetic or orthokinetic aggregation. For perikinetic aggregation:

$$\frac{dN_k}{dt} = \frac{1}{2} \alpha \left\{ \sum_{i=1}^{i=k-1} 2\pi(d_i + d_j)(D_i + D_j)N_iN_j \right\}$$

$$- \alpha N_k \left\{ \sum_{i=1}^{\infty} 2\pi(d_i + d_k)(D_i + D_k)N_i \right\}$$

$$+ \frac{1}{2} \alpha \{8\pi d_k D_k N_k^2\} \tag{27}$$

and for orthokinetic aggregation:

$$\frac{dN_k}{dt} = \frac{1}{2} \alpha \left\{ \sum_{i=1}^{i=k-1} \frac{1}{6}(d_i + d_j)^3 GN_iN_j \right\}$$

$$- \alpha N_k \left\{ \sum_{i=1}^{\infty} \frac{1}{6}(d_i + d_k)^3 GN_i \right\}$$

$$+ \frac{1}{2} \alpha \left\{ \frac{4}{3} d_k^3 GN_k^2 \right\} \tag{28}$$

where D is the effective particle diffusivity, Eq. (16), G is the average shear rate, N is the number concentration of the particles and the subscripts i, j and k define particle sizes where, for collision purposes

$$d_k^3 = d_i^3 + d_j^3 \qquad (29)$$

The effectiveness of collisions in forming aggregates is given by α, and this term may also include corrections for hydrodynamic effects (Sect. 3.1.2). For both Eqs. (27) and (28) the first term on the right hand side describes the formation of particles of diameter d_k by collisions of smaller particles, the second term represents the rate of loss of particles of diameter d_k by collisions. The third term accounts for collisions between like particles in the second term which have been counted twice. It is interesting to note the dominant effect that large particles have on Eq. (28), since these mop up or accumulate a large number of the small particles very quickly during orthokinetic aggregation. For perikinetic aggregation, substituting Eq. (16) into Eq. (27), it is seen that the rate of loss of particles is only a weak function of particle size. The use of Eqs. (27) and (28) with wide particle size distributions is still difficult, but can be made easier by grouping the particles into size bands [68].

The modelling of the growth of isoelectric soya protein precipitates showed some agreement with experimental values over the first few seconds of the growth process and substantial disagreement thereafter (Fig. 16) [99]. An effectiveness factor of 0.24 was used along with a simple maximum size limitation preventing the formation of particles greater than 31 μm diameter, a value obtained by examination of the final size distributions.

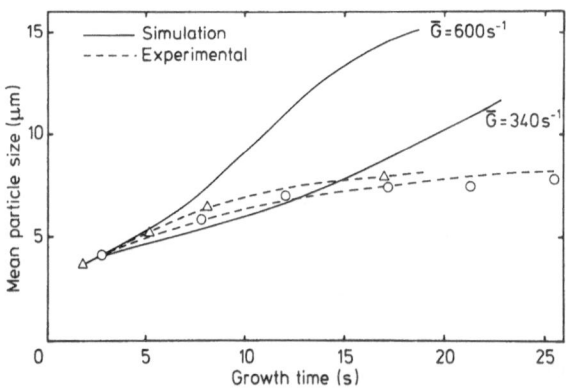

Fig. 16. Growth of soya protein precipitates in a continuous tubular reactor, compared with growth rates predicted by Smoluchowski's orthokinetic aggregation theory, incorporating a maximum particle size limit of 31 μm [99]. Mean velocity gradient, 340 s^{-1} (o); 600 s^{-1} (Δ)

In the precipitation field the growth of calcium carbonate aggregates in a stirred vessel has been successfully modelled using essentially perikinetic aggregation theory with a rate constant orders of magnitude higher than that predicted from the particle diffusivity, see Eq. (16) [100, 101]. Only small precipitate particles of 0.2 to 3 μm diameter were prepared and no allowance was made for shear break-up. Also the shear controlled agglomeration of titanium dioxide particles has been successfully predicted using the same model [102]. The aggregates in these cases are less susceptible to shear break-up than are the relatively large protein precipitate particles. In the flocculation field, the rate of growth of flocs is substantially controlled by floc break-up. Ives and Bhole [103] have modelled the aggregation of spherical particles in laminar shear, simulating break-up by imposing a maximum size limitation on the aggregate formed. Very small degrees of aggregation were studied, only pertinent to the early stages of protein precipitate growth. The computer modelling of floc growth in sheared kaolin-anionic polymer suspensions has also been successful [104]. However, break-up is predicted in terms of an empirical collision effectiveness factor which approaches zero as the experimentally determined equilibrium size is approached. This does have some theoretical justification where larger flocs have lower effectiveness of collisions (Sect. 3.1.2) and also allows for increasingly weak flocs. A theoretical model for floc growth has been proposed by Hunter and Frayne [105]. The growing aggregate is described as an elastic floc structure with a certain yield stress. The break-up of the structure becomes easier with increasing size and this, in conjunction with a modified Smoluchowski theory (allowing for hydrodynamic effects between colliding particles), was used to predict the growth of latex aggregates in laminar shear fields.

The growth of protein precipitates may probably be modelled using orthokinetic aggregation theory along with an empirical equation to describe the rate of shear break-up of the precipitate. However this rate of break-up is a very dominant process as is seen by the rapid divergence between theory and experiment in Fig. 16. Even more significant divergences have been observed over longer ageing times for the growth of casein precipitates [68]. In this case the break-up was based on a maximum size limitation and a series of first order break-up processes determined from the equilibrium particle size distribution. However the time-dependent properties of the protein precipitate cannot be determined with accuracy from the equilibrium data due to the greater strength of the precipitate with increased exposure to shear (Sect. 3.2).

4 Preparation and Ageing of Protein Precipitates

The preparation of protein precipitates involves contacting the soluble protein with a precipitating reagent in a reactor system.

The initial stages of protein precipitation will be governed by perikinetic aggregation forming precipitate particles of a size dependent on the rate of loss of solubility, the rate of aggregation of the protein molecules and the intensity of mixing in the preparation vessel. Orthokinetic aggregation will follow when sufficiently large precipitate particles have been formed. The modelling of protein precipitate preparation should involve population balances [106] along with the normal mass balances and rate equations. However this is not normally possible due to the wide size distribution of precipitate particles commonly obtained from reactors and the inability to incorporate equations describing precipitate break-up (see Sect. 3.2).

In the design of reactor systems important factors that must be considered are the possibility of denaturation due to high localized concentrations of precipitating agent, shear or shear associated damage to protein molecules, foaming or air-entrapment, coprecipitation of proteins. For some proteins the existence of a physico-chemical kinetic controlling step for the formation of nuclei is also important. Processes may be also exothermic and require efficient heat transfer to minimize protein denaturation. However the initial design problem is to establish the requirements of contacting of reactants for nucleation and whether these conflict with the optimum conditions of particle growth.

The basic reactor configurations used for protein precipitation are the batch tank and for continuous operation, the stirred tank reactor (CSTR) and the tubular or plug-flow reactor. All of these reactor types have been extensively investigated in the chemical engineering literature and are considered in several authoritative texts [107, 108, 109]. In this section some of the established principles will be applied to the processes of protein precipitation and ageing.

4.1 Reactor Design

4.1.1 Mixing and Nucleation

As discussed in Sect. 3, the processes of nucleation and particle growth are distinct and controlled by different mechanisms. The factors influencing the time for nucleation are a charge adjustment and/or ion adsorption followed by a Brownian collision. Overbeek [110] gives the time for adjustment of the double layer structure to be of the order of 10^{-8} s. He also indicates that if the double layer equilibrium involves adsorption of potential-determining ions then the surface charge has to be adjusted and the time required for this adjustment may vary from 10^{-6} to as high as 10^4 s. The time for a Brownian collision may be described by the theory of Smoluchowski (see Eqs. (8)–(11)). Assuming that protein precipitation involves some if not all of these processes, and accounting for the fact that a colloidal particle of 0.1 μm will normally incorporate at least 10^3 protein molecules, then the formation of such a particle may take from a fraction of a second up to many minutes. Nucleation times of up to a few seconds are observed for the calcium precipitation of bovine α_s-casein [59] and for the isoelectric precipitation of casein [111]. Other systems, such as the ammonium sulphate salting-out of fumarase [112] and the alcohol precipitation of catalase [22], have been reported to have nucleation times of several minutes.

In the case of fast reactions the major design consideration for the nucleation process is to achieve good mixing to ensure a uniform distribution of precipitating reagent and for efficient heat transfer in the case of exothermic reactions. Although few data exist to define the relationship between the reagent mixing conditions and the subsequent particle size characteristics it might be expected that for very fast protein precipitation reactions the rate of intermixing of reactants will affect the primary particle size and so affect the growth characteristics. Under turbulent mixing conditions, the protein solution may be thought of as being divided into minute packets of fluid defined by the turbulent microscale of mixing (Eq. (20)). The lifetime of these packets depends on the diffusivity of the protein species and the recirculation rates of the fluid through regions of high turbulence. A low molecular weight precipitating agent such as an acid, salt, or organic solvent, will rapidly diffuse into these packets.

The relative rates of protein precipitation and dissipation of proteins from the packets will determine the size of primary protein precipitate particles formed by this mechanism. A similar mechanism for polymeric reactions has been proposed by

Nauman [113]. If the rate of precipitation is very fast with respect to the lifetime of the turbulent eddies, then the basic particle size will be determined by the eddy size and the protein concentration. On the other hand the rate of diffusion of the precipitating reagent to the protein surface or the rate of precipitation may be sufficiently slow that individual groups of protein molecules are insolubilized. These then aggregate readily at a rate controlled 'by perikinetic and subsequently orthokinetic collisions (Sects. 3.1.1 and 3.1.2). Control by the diffusion of the precipitating reagent is particularly relevant to the flocculation of proteins using high molecular weight polymers (Sect. 3.1.3).

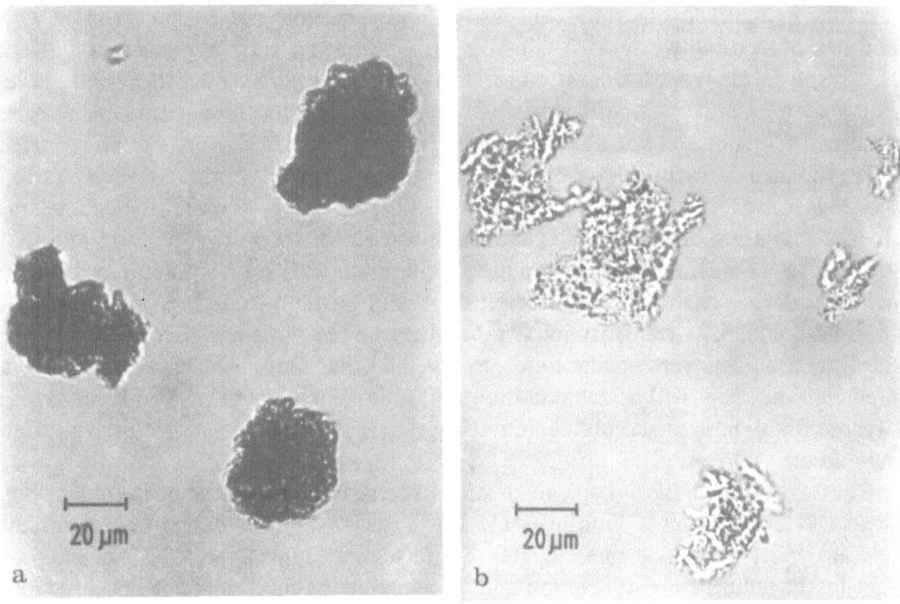

Fig. 17a and **b**. Micrographs of (**a**) isoelectric soya protein precipitate prepared in a batch turbine reactor and (**b**) casein precipitate prepared by salting-out at 1.8 M ammonium sulphate in a continuous stirred tank reactor [114]

The salting-out of casein [67] and the isoelectric point precipitation of soya protein [114], both in turbulent mixing vessels, show the formation of basic particles of approximately 1 µm diameter (Fig. 17). The precipitating reagents used are both of low molecular weight and the formation of these basic particles is probably accounted for by the rapid diffusion of the reagent into the fluid packets of protein.

When the nucleation times are of the order of tens of seconds or more a key choice relates to the selection of a batch or continuous reactor system. Batch vessels should be designed to avoid potential denaturation effects or overprecipitation. The influence of batch mixing conditions and acid type on soya protein structure during isoelectric precipitation has been reported by Salt et al. [20]. They found that the extent of protein modification was dependent on the acid anion but that for this system poor mixing did not cause major damage since the reaction is very fast. In a continuous process the retention time is important and the yield of precipitate may be affected by the precise conditions of mixing especially if short circuiting occurs. In this respect plug flow reactors may be more suitable since ideally the fluid elements have the same retention time.

The influence of the contacting conditions was highlighted by Foster et al. [112] who showed that they have a significant effect on the salt concentration at which precipitation occurs for the salting-out of alcohol dehydrogenase and fumarase from clarified yeast homogenate with ammonium sulphate. The data for fumarase is shown in Fig. 18 with different contacting procedures producing a shift in the salting-out curves. Foster et al. suggested that the shift in the curves could be due to differences in the localized pH conditions with, for batch contacting, solid ammonium sulphate resulting in higher localized concentrations (and hence lower pH) than obtained with addition of a partially saturated salt solution. A variation in the degree of mixing will also result in different localized concentrations and so affect the salting-out curve. This may account for the difference between batch and continuous processing. The influence of pH on the salting-out curve is discussed in Sect. 2. Foster et al. reported that a similar shift in the salting-out curve was observed in the continuous precipitation of β-galactosidase from *Escherichia coli* [112]. They also observed that the continuous flow precipitation of ADH and fumarase produced the largest particle size and that in the batch contacting experiments the precipitation with solid ammonium sulphate produced larger particles than precipitation with partially saturated salt solutions.

Good mixing is particularly important when the reactants are viscous fluids since the formation of layers of one fluid next to the other may result in a non-uniform precipitation. This will affect continuous fractional precipitation processes where overprecipitation is unacceptable and also processes where irreversible precipitation may occur.

Control of the submicroscopic nucleation process by effective mixing is not feasible, since it occurs within a limiting scale below which hydrodynamic forces do not operate [115]. However a small degree of nucleation control may be achieved by varying the temperature of reaction and so changing the rates of molecular diffusion.

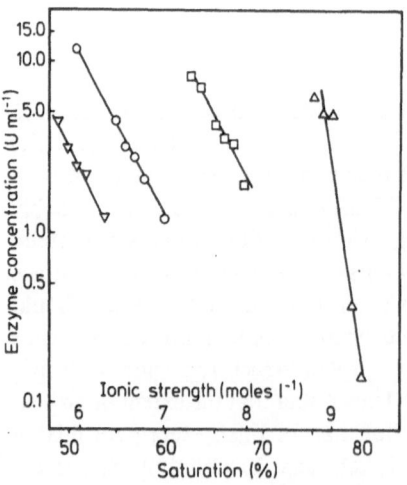

Fig. 18. The influence of contacting procedure on the salting-out curve for fumarase [112]. ▽, ○-batch contacting; □, △-continuous contacting, residence time 96 s in 0.38 l C.S.T.R. and 918 s in 1.87 l C.S.T.R., respectively; ▽-ammonium sulphate solid, ○,□,△-saturated ammonium sulphate solution

Figure 19 illustrates this effect for isoelectric casein precipitation with an increase in temperature of 30 °K decreasing the apparent time of precipitation by about 3.5 s.

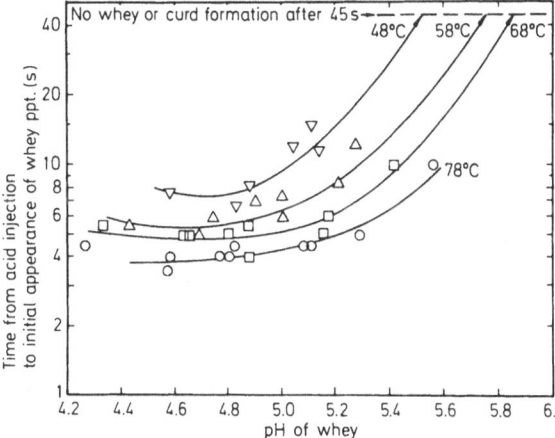

Fig. 19. Effect of temperature of coagulation and pH of whey on the time taken from acid injection to the initial appearance of whey precipitate. Coagulation temperature: o, 78 °C; □, 68 °C; △, 58 °C; ▽, 48 °C; initial pH of milk: 6.4 [111]

Good mixing is usually achieved by incorporating a zone of turbulence such as that of the impeller zone in a stirred tank, by jet mixing or by the use of a static mixing device or recycle in a continuous tubular reactor. The turbulent process is used to break up the fluid elements to some limiting size at which viscous forces prevent the turbulent motion and molecular diffusion becomes the controlling factor. This limiting size [108] is usually called the Kolmogorov length, η, and is defined by Eq. (20) in Sect. 3.1.2. Molecular diffusion reduces the intensity of segregation of the fluid elements with or without turbulence but the presence of turbulence breaks the fluid into smaller elements and so increases the surface area for diffusion.

A complete model of mixing in any reactor system must define both the microscale (mixing of individual molecules) and the macroscale processes (mixing of groups or packets of molecules) [113]. 'Maximum mixedness' and complete segregation represent the opposite extremes of mixing. The condition of maximum mixedness corresponds to the maximum amount of molecular-level mixing possible with a given residence time distribution. Complete segregation occurs when the fluid elements experience the bulk flow patterns of the reactor without molecular interchange. Typically fluid enters a continuous reactor in the segregated state and leaves in a condition of maximum mixedness. For slow precipitation reactions carried out continuously the presence of segregation may limit the interaction of reactants for a significant time and so lower the yield if the reactor residence time is short.

Since the level of micromixing can be important in the protein precipitation process the use of a criterion such as a segregation number will enable the state of micromixing to be determined. Nauman [113] describes the diffusion process between fluid elements by a droplet diffusion model and defines the segregation number by:

$$N_{seg} = R^2/\pi^2 D \bar{t} \qquad (30)$$

where R is the droplet size and can be considered as the turbulent microscale, η; D, diffusivity between droplet and bulk fluid; $\bar{t}$, reactor mean residence time. For a value of $N_{seg} \gg 1$ the reactor is completely segregated and for a value of $N_{seg} \ll 1$ it is in a state of maximum mixedness. The effect of segregation can be important for systems when the feed streams enter the vessel separately and with high viscosities or low diffusivities, and may account for some of the discrepancies observed by Foster et al. [112] for precipitation in different reactor systems (Fig. 18). Segregation numbers for the CSTRs in this figure were of the order of 0.5–5, corresponding to some degree of segregation and support Foster's proposal that the shift in salting-out curves was due to contacting conditions.

An alternative model considers the intensity of segregation, I_s defined in terms of concentration fluctuations at a point and described by the equation [116]:

$$I_s = \frac{\overline{C_A' C_B'}}{C_{A_0} C_{B_0}} \tag{31}$$

where C_A and C_B are the average concentration fractions of components A and B and $\overline{C_A' C_B'}$ is the time average value of the product of the instantaneous fluctuations of the concentration fractions; subscript o refers to the initial state. In this case, $I_s = 1$ represents complete segregation (no mixing) and $I_s = 0$ represents uniform mixing (zero fluctuations). Figure 20 indicates the influence of impeller type, feed rate and feed location on the levels of segregation in stirred tanks, showing that the use of the lower shear mixing devices to minimize the possibility of shear denaturation may result in significant segregation effects [117].

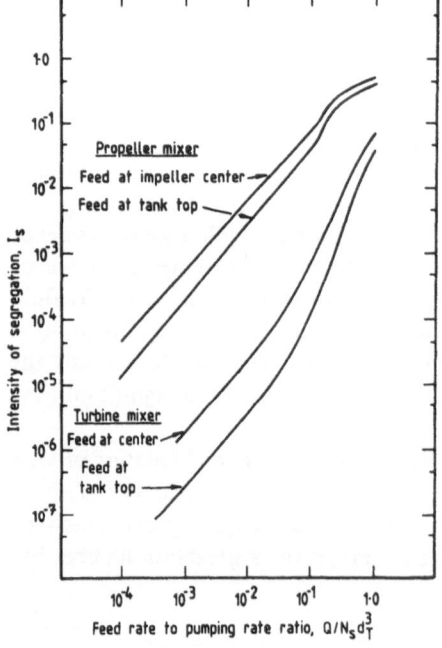

Fig. 20. Levels of segregation at the discharge point for tanks stirred by Rushton turbines and propellers. Turbines and propellers centred in tank; impeller diameter, d_T, one-third of tank diameter; $Q/N_s d_T^3$ is a measure of feed rate to pumping rate ratio [117]

Parameters such as the segregation number or the intensity of segregation should assist in the design of precipitation reactors and highlight the importance of a knowledge of the precipitation kinetics in the selection of reactor configurations. Bench scale equilibrium studies used to determine precipitation characteristics such as the salting-out curves (Fig. 3) are alone inadequate for a realistic approach to industrial scale-up.

The detailed distribution of energy in a stirred vessel is not clearly defined and so the use of parameters, such as the average power/unit volume, are not totally satisfactory for scale-up. When the effects of shear denaturation of the soluble protein or shear disruption of the formed precipitates are of primary importance the conditions of mixing in the impeller zone where the maximum shear stresses occur, will form a design basis. The high shear stresses associated with the trailing vortex system produced by turbine impellers have been described by Van't Riet and Smith[118]. Mixing with lower maximum shear stresses can be achieved with propellers or helical ribbon type impellers. However use of these impellers is usually accompanied by higher recirculation rates which can result in denaturation due to the higher rate of reformation of air/liquid interfaces in a non-flooded reactor. Many of these problems can be overcome in sealed continuous systems. Vrale and Jordan[119] investigated the use of back-mix (CSTR) and plug-flow reactors for rapid mixing in the water treatment field involving colloid destabilization (or precipitation) of metal hydroxide species, and found that the back-mix reactor was inferior for rapid mixing. Bratby[120] considered the design of such devices on the basis of mean velocity gradients. In systems where the contacting of reactants is not critical then the conditions of mixing can be defined by the conditions for optimum particle growth.

4.1.2 Particle Ageing

Having formed the nuclei the objectives of any subsequent treatment are to form large, dense, stable aggregates which can be readily recovered by centrifugation. The growth of precipitated aggregates (larger than about 0.1–1 μm) is a shear controlled aggregative process as described in Sect. 3.1.2, with the history of preparation influencing the size, overall shape and degree of compactness of these aggregates. Even though the precise growth kinetics have not been defined for protein systems the manipulation of concentration and shear rate has been shown to result in an advantageous variation in the particle size distribution for isoelectric soya protein precipitates in batch[89] and continuous tubular reactors[99] and for ammonium sulphate precipitated casein in a CSTR[67]. The growth of soya protein isoelectric point precipitate in a tubular reactor shows increased rates of growth at higher average rates of shear and greater protein concentrations[99]. Both of these results would be expected from orthokinetic controlled aggregation. However precipitate prepared at a higher rate of shear tends to a smaller final particle size due to the greater rates of shear induced break-up. That is, neglecting any irreversibility of precipitate break-up, there is a shift in the balance between aggregation controlled growth and shear controlled break-up. A similar shift occurs when the protein concentration is reduced, again smaller particles being obtained at the end of the process. The effect of the rate of shear is further emphasized in Fig. 21 for the ageing of casein precipitate prepared by salting-out[67]. The lowest average rates of shear give substantially the

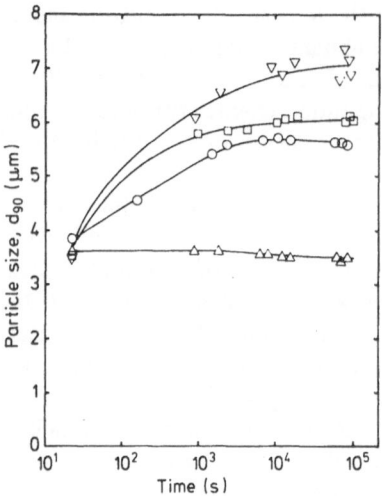

Fig. 21. Growth during ageing in a helical ribbon stirred reactor of casein precipitates prepared by salting-out with 1.8 M ammonium sulphate [67]. Mean velocity gradients in the reactor, calculated on the basis of power dissipation per unit volume, are; $\triangledown$, $12\,s^{-1}$; $\square$, $31\,s^{-1}$; $\circ$, $51\,s^{-1}$; $\triangle$, $154\,s^{-1}$

largest final precipitate particle size, the shapes of the precipitate growth curves again supporting the idea of a balanced process between growth and break-up. The scale-up of this aggregation process is particularly difficult for stirred tanks since all the shear stresses cannot be simultaneously controlled, hence it is convenient to consider the stirred tank as consisting of two mixing zones [108], the high shear impeller zone and the lower shear bulk fluid. If the critical zone can be identified then the ageing process might be considered as a function of the intensity of shear in that zone and the time of exposure, i.e. Gt or $\bar{G}$t. This approach is taken in the water treatment field for the formation of flocs in stirred tanks, where the growth is considered to be related to the bulk conditions and the product of the mean velocity gradient, $\bar{G}$, and exposure

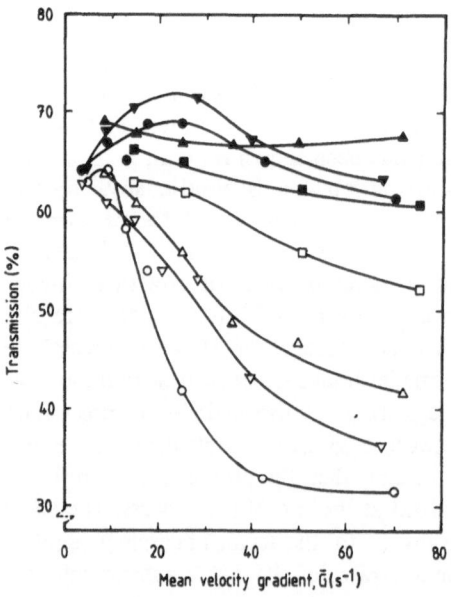

Fig. 22. Coagulation-flocculation efficiency as represented by transmission after a given settling time, as a function of mean velocity gradient, $\bar{G}$, in a Couette flocculator and for various types of impellers in a square tank [122]. Settling time: open symbols, 300 s; closed symbol, 1800 s. Square tank impellers: $\circ$, $\bullet$, turbine: $\triangledown$, $\blacktriangledown$, paddle: $\triangle$, $\blacktriangle$, propeller. Couette flocculator: $\square$, $\blacksquare$

time, t, is known as the Camp number [121]. Values of $\bar{G}t$ of 10^4 to 10^5 with values of $\bar{G}$ from 10 to about 100 s^{-1} are recommended in the water treatment field. The geometry of both the tank and the impeller are critical and comparisons between different tanks can be confusing due to the different shear distributions. The impeller type is of particular importance since different configurations can affect both the time of exposure and maximum level of shear in the impeller zone. Leentvaar and Ywema [122] showed the wide variation in flocculation efficiency of a water treatment system as a function of mean velocity gradient and impeller type (Fig. 22). They and others [103] also indicated the advantages in having higher shear rates in some parts of the tanks to enhance the overall flocculation performance. Similar effects were observed by Virkar et al. [99] for the continuous precipitation of isoelectric soya protein in a tubular reactor (Fig. 23). At a constant flowrate the aggregation performance was improved by the insertion of grid turbulence promoters at regular intervals down the reactor. This might be seen as an approach to a stirred tank configuration where the grids correspond to the high shear impeller zone. Other studies with tubular systems have shown the significance of helical configurations in enhancing aggregation under laminar flow conditions [72].

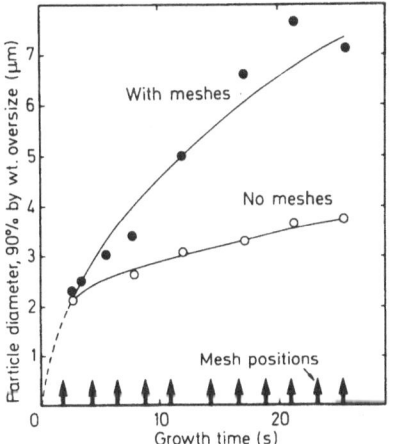

Fig. 23. Effect of turbulence promoters on isoelectric soya protein precipitation in a continuous tubular reactor [99]. Protein concentration: 2.0 kg m^{-3}; average fluid velocity: 0.47 m s^{-1}

The parameter $\bar{G}t$ is also found to correlate the strength of isoelectric soya precipitate prepared in a batch reactor [92]. Figure 15 shows the change in the mean size of the particles after subjecting the sample to a capillary rate of shear of 1.7×10^4 s^{-1} for 0.065 s, as a function of $\bar{G}t$. These conditions of shear are considered to be similar to those encountered in high speed rotating equipment. The improved aggregate strength at $\bar{G}t$ values of about 10^5 is attributed to an infilling/rearrangement mechanism resulting in closer, more stable, packing. This effect may be influenced by the reactor configuration and the existence of a distribution of velocity gradients may be the main cause of increased strength. The improvement in strength of flocs with ageing has also been noted by Kitchener and Gochin [123] as being due to the preferential survival of tightly packed nodular flocs in water treatment systems. Comparison of the strength of particles prepared in a batch reactor and a continuous tubular reactor shows significant differences [124]. Figure 24 gives the volume of fine particles produced as a result of subjecting a sample to a capillary rate of shear of 9×10^4 s^{-1} for varying times. The continuous tubular reactor precipitate exhibits significantly less resistance to disruption with the volume of fines increasing from about 2.5 % to over 25 % for shear exposure times greater than about 0.05 s. Subsequent ageing of the tubular reactor-prepared precipitate in a batch tank only resulted in a small improvement in the aggregate strength. This shows that the precise conditions

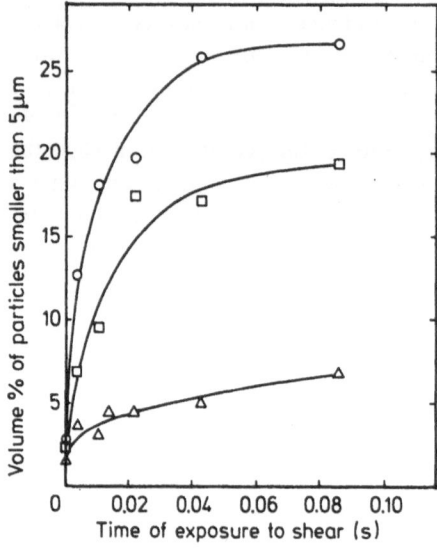

Fig. 24. The effect of preparation conditions on the resistance of isoelectric soya protein precipitate to short duration shear break-up [124]. Average rate of shear; 9×10^4 s^{-1}; protein concentration: 25 kg m^{-3}; preparation conditions: o , tubular reactor, $\bar{G} = 340$ s^{-1}, $\bar{G}t = 8.5 \times 10^3$; □ , tubular reactor, $\bar{G} = 340$ s^{-1}, $\bar{G}t = 8.5 \times 10^3$, plus 2400 s batch ageing at $\bar{G} = 200$ s^{-1}; ▵ , batch stirred tank, $\bar{G} = 200$ s^{-1}, $\bar{G}t = 3.6 \times 10^5$

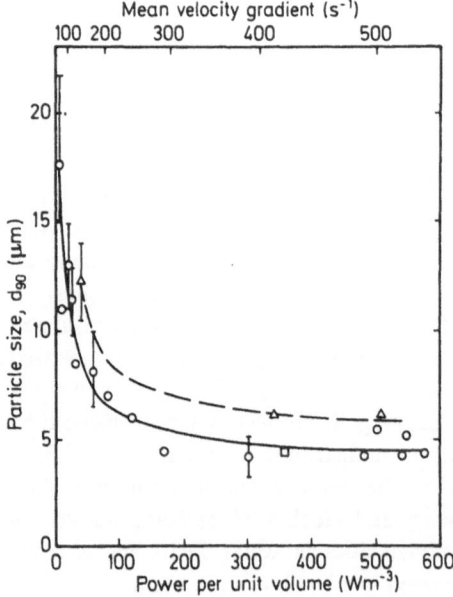

Fig. 25. The relationship between aggregate size and power per unit volume for the preparation of isoelectric soya protein precipitate in batch stirred tanks [89]. Protein concentration: 30 kg m^{-3}; vessel volumes: ▵ , 0.27 l; o , 0.67 l; □ , 200 l. Ageing time, 600 s

of mixing determine the strength and highlight the difficulties that may be encountered in selecting a suitable reactor design. It also points to the possibility of significant changes in aggregate strength resulting from scale-up.

Scale-up of batch reactors in relation to the particle size distribution on the basis of dissipated power per unit volume is shown for the same isoelectric soya protein precipitate in Fig. 25 [89]. The reasonable agreement for the d_{90} values (i.e. the fine end of the distribution) suggests that this parameter may be useful in scale-up for subsequent centrifugal recovery, since the recovery efficiency is determined by the volume of fine particles (Sect. 5). Since a long time may be required to achieve significant particle growth, the choice of reactor configuration can be important for the ageing process. O'Melia [125] compares the flocculation performance of a plug flow or batch reactor with that of a CSTR. He describes the aggregation rate for a homogeneous colloid as a pseudo-first order reaction (Eq. (18), Sect. 3.1.2). For a batch or plug-flow reactor the change in particle numbers due to flocculation is given by:

$$\ln \frac{\bar{N}_0}{\bar{N}_t} = \frac{4}{\pi} \alpha \varphi_v \bar{G} t_{PFR} \tag{32}$$

where $\bar{N}_0$ is the initial number of particles and $\bar{N}_t$ is the number of particles after time $t = \bar{t}_{PFR}$.

For m equal size CSTRs:

$$t_{CSTR} = \frac{\pi}{4} \frac{m}{\alpha \varphi_v \bar{G}} \left[\left(\frac{\bar{N}_0}{\bar{N}_t} \right)^{1/m} - 1 \right] \tag{33}$$

where t_{CSTR} is the total residence time in the m tanks. Comparison of these reactors for a value of $\bar{N}_t/\bar{N}_0 = 0.01$ gives a single CSTR volume of 22 times that for a plug-flow reactor. Three CSTR's in series require a volume only 2.4 times that of a plug flow reactor. Application of these equations to the aggregative ageing of protein precipitates depends on the accuracy of Eq (18) and the assumption that the collision effectiveness factor, α, is independent of size. It also assumes that at some stage in the protein precipitation process, such as immediately after mixing, the precipitate can be considered as a dispersion of homogeneous primary particles. These assumptions are simplistic but the equations indicate a method by which batch laboratory data, in the form of growth kinetics, may be used to design large scale continuous precipitate ageing systems. These principles are incorporated in a recent patent [126] in which promotion of particle growth in flowing suspension is achieved in a continuous flow-through chamber. Internal partitions cause vertical flow and each section has a greater flow cross-section than the preceding upstream section resulting in a reducing velocity gradient.

4.2 Acoustic Conditioning

Processes that require extensive ageing times to achieve acceptable recovery in terms of both clarification and solids dryness are often restricted to batch operation. The demand for continuous operation has resulted in the application of conditioning devices to improve the sedimentation and dewatering characteristics. An early patent by Hiedemann [127] described the treatment of liquids containing suspended particles with 'sound and ultrasound' waves. Several applications of the principles described in this early patent have been reported for inorganic chemical systems [128, 129].

The first reported application to protein systems was by Jewett [130] who observed an increase in the sedimentation rate of antihemophilic factor, precipitated by PEG, when two vibrating sources were imposed on the dispersion at different frequencies. Industrial scale application of this technology is reported by Foster [131] for the conditioning of plasma protein precipitates in a continuous flow Cohn fractionation process. Application of this technology to other protein systems has been limited by the lack of understanding of the mechanisms involved.

5 Centrifugal Separation

The preceding sections have considered the methods of protein precipitation and the design of reactor systems. Due to the widespread use of protein precipitation and the high operating and capital costs of the solid-liquid separation equipment involved the efficiency of the separation step is central to product recovery. The protein source and precipitation method are significant factors in determining the separation efficiency and in defining the type of separation equipment to be used. The objectives are for both a well clarified supernatant and a solid of maximum dryness, giving high separation efficiency and minimizing the subsequent processing costs. The fundamental mechanical methods of achieving solid-liquid separation are: the use of pressure forces in straining, pressing or filtration; gravitational or buoyant forces in sedimentation or flotation; and centrifugal forces in centrifugal sedimentation. There is a wide variety of such equipment much of it initially designed to meet the needs of the chemical industry. Performance predictions depend on a knowledge of specific features of both the solid and liquid phases. The colloidal and gelatinous nature of many protein precipitates causes blockage of cloths or meshes and the formation of compressible sludges and so precludes the selection of equipment using filtering mechanisms unless a filter aid is incorporated. Use of filter-aid may result in product contamination and in many processes such as the separation of pure albumin or γ-globulin fractions from blood plasma [132], the adsorption effects of the filter-aid prevent acceptable recoveries. Other disadvantages associated with some filtration equipment are the difficulties of containment due to equipment size, the relatively long preparation and start-up times and the occurrence of foaming in vacuum systems. However, in some systems the particular nature of the materials and process requirements make separation by filtration a suitable method. Examples include the use of rotating discharge pre-coat filters for the coarse filtration of hot ethanol precipitated albumin [133] and for beer clarification. The removal of haze forming protein-polyphenol precipitates in beer is also widely carried out by plate-and-frame, pressure leaf and candle filters [134]. Other examples of filter press use are described by Nyiri [135] for the recovery of ammonium sulphate precipitated pectolytic enzyme, and by Underkofler (in Reed, 1975) [136] for the recovery of an extracellular bacterial amylase from clarified broth by precipitation with denatured alcohol. Rotary vacuum and rotary drum pressure filtration are also used when the solids nature is suitable. Detailed descriptions of the selection and operation of rotary drum filters can be found elsewhere [137].

Filtration can also take place in centrifuges where the centrifugal field acts as the pressure force. The principle of these perforate bowl machines is shown in Fig. 26f. However, the recovery of proteins by this method is restricted to batch type basket or peeler centrifuges because of the small size of the precipitate particles. Their use is also limited due to the compressible nature of precipitated protein which can result in severe blinding of the perforations at the high filtration driving force. Filtering centrifuges are only suited to high value products due to the substantial relative cost per unit filtration area.

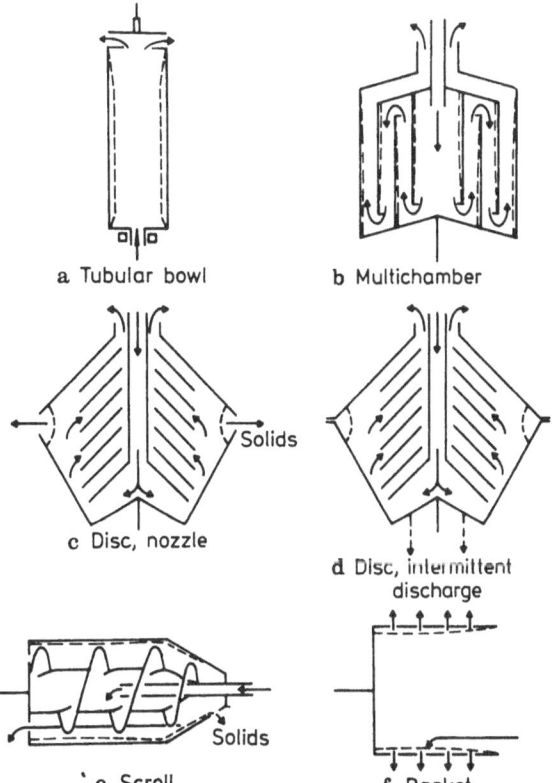

a Tubular bowl   b Multichamber

c Disc, nozzle

d Disc, intermittent discharge

e Scroll

f Basket

Fig. 26a—f. Centrifuge configurations. Arrows show the path taken by the liquid phase [144]

Other methods of solid-liquid separation involving non-centrifugal forces include the use of foam separation methods [138] such as dissolved air flotation for the recovery of isoelectric casein [139], and the use of screw presses for the dewatering of casein curds [140]. Recent publications have shown the feasibility of the application of ultrafiltration or cross-flow filtration for the concentration of particulate systems such as cells and debris from fermentation broths [141], and leaf proteins [142]. Gravity sedimentation using lamellar settlers has also been shown to be feasible for microbial cells [143] and in some cases may be appropriate for the separation or preconcentration of precipitated protein.

The principal approach to the recovery of protein precipitates is centrifugal sedimentation and this method is considered in detail here. It is based on the existence of a density difference between the particulate solids and the suspending liquor, and the separation of the solids by the application of a centrifugal force. The apparent advantages over filtration are the low operating cost, the lack of contamination (since filter-aids are not necessary), the brief preparation and start-up times required, and the low space requirements. Other possible advantages such as containment, limited operator demand, continuous operation, and ease of automation depend on the process requirements and the particular machine characteristics. The specific features of protein precipitates are of particular importance in the selection and operation of centrifuges since in many cases they have resulted in the requirement for special modifications to achieve acceptable separation.

5.1 Special Features of Protein Precipitate Separation

The special features of protein precipitate separation are associated with both the nature of the solid and liquid phases as well as product specification requirements. Most processes require sanitary operation where the equipment must meet the relevant standards of design such as construction in food grade materials, surface polish and lack of void spaces as well as having clean-in-place facilities. In some processes the equipment must also be capable of sterilization. This can create considerable difficulties in high speed rotating machines where the presence of seals provides a source of contamination. The safe operation of centrifuges can also be a major problem since many new processes involve the fractionation of proteins from fermentation liquors where a level of biologically toxic materials persists through several processing stages. The rotating machinery can disperse these substances in aerosol form, causing an environmental hazard. Most existing centrifuges display this problem and without expensive modifications the only alternative is to contain the equipment in some form of cabinet as discussed by Wang et al. [144]. The solid phase may also create biosafety problems, particularly for batch machines where the solids are handled by process operators. Equally the use of sealed systems and protective clothing can result in considerable operational difficulties. The processing of systems with organic solvents can create explosive hazards and inert gas blanketing may be required [145]. Another safety aspect often overlooked is the corrosive nature of precipitants such as ammonium sulphate.

Apart from the mechanical operating problems, the main difficulties in protein precipitate solid-liquid separation are associated with the physico-chemical nature of the protein system. The nature of proteins can lead to foaming and air entrapment resulting in low effective settling velocities for centrifugation as well as protein denaturation due to the presence of large air/liquid interfaces. Denaturation may also occur due to a combination of temperature — and interface associated — shear effects as discussed in Sect. 4, although such effects would be limited primarily to the soluble protein and would depend on the concentration and sensitivity of the particular protein [146, 228]. Temperature control is a critical parameter in many processes such as the cold-ethanol fractionation of blood plasma [147]. Control to ± 0.5 °C is required as well as the ability to operate at low temperature levels. Temperature rises on passing through the bowls of uncooled centrifuges are excessive due to the low flowrates used for difficult protein systems and have resulted in the introduction of cooled machines. The major part of the temperature rise can be attributed to the effects of windage on the centrifuge bowl [148]. Other limitations arise from the small particle size, the low density difference between the precipitate particle and the associated liquor, the high liquid viscosity and the fragile nature of the aggregated particle. If these problems are overcome and the precipitate is sedimented then the complex rheological behaviour of the sludge often presents a problem in the operation of automatic discharge mechanisms.

Particle size determination of protein precipitates is usually performed by conductance or sedimentation analysis. Microscopic examination techniques are laborious to operate so as to yield statistically meaningful data. The variation in conductivity across an orifice immersed in an electrolyte is a function of the volume of a particle passing through the orifice [149]. The heterogeneous structure of a protein precipitate aggregate raises several questions about its conductivity. Free electrolyte in totally open pores will be conductive whereas free electrolyte in a pore closed at one or both ends within the aggregate will be non-conductive. On the basis of particle size analysis using different techniques it has been estimated that the conductometric method accounts for at least 93% of the diameter of 6 micrometre isoelectric point soya protein precipitate particles [114]. Sedimentation techniques involve several unknowns in the equation used for their analysis due to the difficulty of defining the density and shape of the precipitate particles. Particle sizes vary according to the precipitation system and also with the extent of ageing, as discussed in Sects. 3 and 4 but typical sizes range from about 0.1 to 50 μm. Only the small particles (0.1 to 5 μm) present sedimentation difficulties in high speed industrial centrifuges.

The density difference is often very low due to the precipitants used. Typical ammonium sulphate concentrations of 25% to 75% saturation give liquor densities of 1070 to 1190 kg m^{-3} and polyethyleneglycol liquor densities are about 1060 kg m^{-3} at concentrations of about 20%. Maximum crystalline protein densities are of the order of 1350 to 1400 kg m^{-3} [150]. Densities of the solvated proteins range from 1280 to 1366 kg m^{-3}. Buoyant densities of isoelectric point precipitates of the globular proteins ovalbumin and bovine serum mercaptalbumin are 1292 and 1282 kg m^{-3}, only 6–9 kg m^{-3} greater than that for the soluble proteins [151]. However the existence of interstitial liquor in the aggregated precipitate results in typical hydrodynamic densities of about 1050 to 1250 kg m^{-3}.

The density of protein precipitate aggregates is a function of the amount of entrapped liquor, the density decreasing sharply from that of a basic particle to a relatively steady or slowly decreasing value for large aggregates. The basic particle buoyant density for isoelectric soya protein precipitate is approximately 1296 kg m^{-3}, the aggregate density, ϱ_a (kg m^{-3}), being related to aggregate diameter, d (μm), as follows [114]:

$$\varrho_a = 1004 + 246d^{-0.408} \tag{34}$$

$$1 < d < 35$$

Similarly for casein precipitate prepared by salting-out with 1.8 M ammonium sulphate:

$$\varrho_a = 1136 + 31d^{-0.441} \qquad (35)$$

$$1 < d < 27$$

These are similar relationships in form to those measured for clay-aluminium flocs [104].

For the theoretical description of a floc produced by the successive random addition of particles [152]:

$$\varrho_a = \varrho_f + k_D d^{-0.676} \qquad (36)$$

The smaller density power functions in Eqs. (34) and (35) indicate a tighter packing of the protein precipitate aggregate. However comparison with a model based on cubic packing of 1 µm diameter spherical particles (Fig. 27) shows a higher level of interstitial water for the larger protein precipitate particles. This may be due to particle non-sphericity, interaction of protein surface groups and also more random packing. The higher voidage of the casein precipitate compared to that of isoelectric soya is supported by microscopic examination of the two forms of precipitate aggregate (Fig. 17), and the shape of the profile can be expected to change depending on the particular protein, the shear history and the method of precipitation.

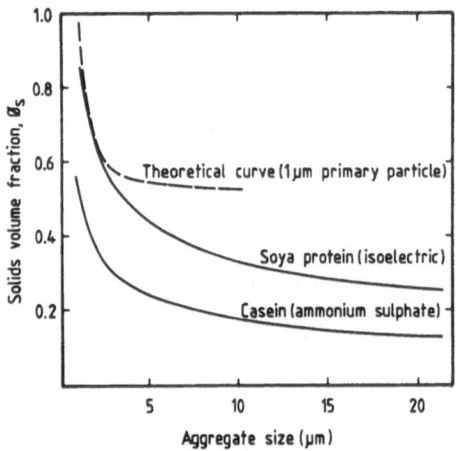

Fig. 27. Solids volume fraction (φ_s) of isoelectric soya protein precipitate and ammonium sulphate precipitated casein as a function of aggregate size [114]. – – – – theoretical curve for cubic packing of 1 µm diameter particles

The effect of the method of precipitation on the precipitate density is shown in a study of casein curds [153]. Lactic acid precipitation gives a density of the curd without free water of 1360 kg m^{-3} while sulphuric acid preparation tends to give higher densities of 1370 to 1400 kg m^{-3}. Curds prepared using rennet show even higher values of 1440 to 1470 kg m^{-3}. The study also emphasizes the problem of air entrainment during precipitation which can yield precipitates with a density less than that of the surrounding medium. This may also occur during the salting-out of proteins where the high salt concentration causes dissolved gases to come out of solution [67].

The viscosity of protein suspensions is often high and in many cases this is due to the presence of non-protein components such as β-glucans in green beer or nucleic acids in fermentation product streams. Removal of nucleic acids by manganous sulphate and heat treatment has been shown to result in a reduction in the viscosity of disrupted bacterial suspensions and so facilitate subsequent protein fractionation [154]. Other methods of removal include precipitation by metal ion [155], polyethylene-imine [156] and streptomycin sulphate [157] or alternatively by the action of added nucleases [158] such as deoxyribonuclease [159]. Reduction of the viscosity can also be achieved by increasing the operating temperature, although in many systems this is not an acceptable alternative.

The strength of the aggregated precipitate and its ability to withstand the intense levels of shear associated with high speed rotating equipment is crucial to the efficient performance of centrifuges since the production of fine particles by shear break-up will result in poor clarification. Most of the proposed break-up models discussed in Sect. 3.2 have been developed for the water treatment field, and so the work has been directed towards systems subject to gravitational sedimentation. Experimental study of high shear particle break-up of protein precipitates has shown it to be critically dependent on the time of exposure. Only a small increase in the number of fine particles will result from short time exposure to shear of isoelectric soya protein precipitate [92, 89]. However, for other protein systems significant numbers of fine particles may be produced due to the different protein-protein interactions involved (Sects. 2 and 3). The critical size at which a particle will not be recovered is defined by the particular combination of protein and centrifuge characteristics as described in the following section. Vinogradova et al. [160] gave an analysis of the disruption of amorphous particles and derived relationships for the limiting size for particles which are not broken up in a centrifugal field.

Having sedimented the precipitate the problems of recovery are related to the complex rheology of the protein sludge and its effect on the operation of intermittent and continuously discharging machines. Protein sludges are sticky, gel-like materials and often exhibit viscoelastic properties. They may display time dependent behaviour and be thixotropic or rheopectic by the occurrence of structural breakdown or formation. Detailed studies of the effect of the rheological characteristics on centrifuge performance are rare, although there is extensive literature on the rheology of both dilute [161] and concentrated dispersions [162] with considerable advances being made in the theoretical description of dispersion flow properties [163, 164, 165, 166]. Many protein precipitates will not dewater readily, resulting in wet sludges which are contaminated with significant quantities of supernatant and so require a series of washing steps which contribute to product loss through low efficiency and denaturation. Of particular interest to the operation of continuous discharging centrifuges is the occurrence of shear thickening behaviour in concentrated dispersions [167]. At a critical concentration and shear rate the rheological characteristics of many dispersions change from shear thinning to shear thickening. This may account for the difficulties found in ensuring a smooth continuous discharge of concentrated protein sludges from scroll discharge centrifuges. The shear thickening of soya protein isolate dispersion was shown by Lee and Rha [168] and was observed by Bell and

Dunnill [124] for the recovery of isoelectric soya precipitate in a pilot scale scroll discharge centrifuge. The shear dependence of the cake viscosities of blood plasma fractions II + III and V from a pilot plant decanter centrifuge were reported by Meier [169]. The rapidly increasing viscosities at low shear rates precluded the operation of the scroll centrifuge at low differential speeds, which were necessary to minimize turbulence and so achieve satisfactory clarification. The cohesive properties and hardness (the compression force required for a 70% deformation) of centrifugally sedimented soya protein precipitates were investigated by Lee and Rha [170], who found that the type of precipitation reagent resulted in different properties at low temperatures, with the isoelectric precipitate being less cohesive than the calcium precipitate. However, this difference was found to become negligible at high temperatures. They found also that the hardness of the sediment produced by isoelectric precipitation increased rapidly, whereas that produced by calcium precipitation responded slowly to the increasing centrifugal force, (Fig. 28). Vu and Munro [171] have reported the effect of pressure on the dewatering of casein curds and shown a critical pressure to result in surface sealing. This effect explains the observation of better dewatering of casein obtained in decanter centrifuges compared to screw presses.

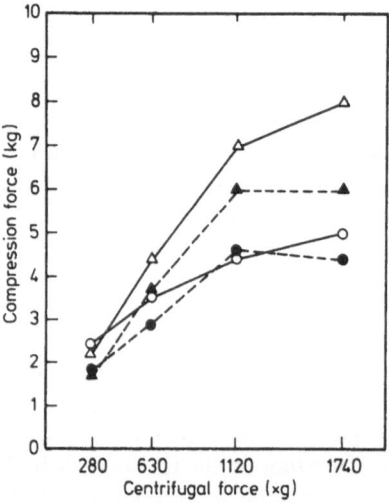

Fig. 28. The relationship between the hardness of precipitated soya protein, as defined by the compressive force required for a 70% deformation of the sample, and the centrifugal force applied in separation [170]. Application time of centrifugal force = 600 s. Isoelectric precipitation, △, 25 °C; ▲, 80 °C. Calcium precipitation, o, 25 °C; ●, 80 °C

These examples clearly indicate the significance of the particular system properties on the settling and dewatering characteristics of precipitated soya protein. Both the sedimentation characteristics and the sludge flow properties are influenced by pH, temperature, ionic strength, precipitating reagent, and in some cases the conditions of pretreatment of the raw materials such as the extraction solvents [172]. The addition of reagents to improve the product quality can have detrimental effects on the dewatering characteristics of the sedimented sludge. For example, sodium metabisulphate in the production of leaf protein isolate [173] resulted in a decrease in curd size and firmness. At the same time it was found to reduce the quantity of fibrous curds which had previously resulted in blockage of centrifuge ports.

5.2 Principles of Centrifugal Separation

The rate of particle sedimentation is defined by the balance between the centrifugal force, the Brownian diffusional force and the kinetic drag of the fluid. For an isolated spherical particle in an infinite fluid medium with negligible Brownian diffusional forces, this leads to Stokes law, where the equilibrium settling velocity, v_0, is given by:

$$v_0 = \frac{d^2 \Delta \varrho}{18\mu} \omega^2 r \tag{37}$$

Application of this theory to centrifugal sedimentation has been considered in detail by many authors [174, 175, 176, 177]. This description is of course simplistic and gives an estimate for particles with low settling Reynolds numbers, $Re_p < 0.2$. Corrections must be made to account for the effects of flow in narrow channels where a particle size of 10% of the gap may result in particle-wall interactions. Non-equilibrium settling may also occur where the proximity of the sedimentation surface is such that larger particles do not achieve the Stokes velocity v_0. However, since the separation efficiency is affected by the behaviour of the smallest particle, non-equilibrium effects are usually not important to the overall efficiency. Particle non-sphericity can have a significant effect resulting in a reduced sedimentation velocity even for slight surface roughness. Another significant factor is the effect of hindered settling. In this case the sedimentation velocity can be modified by an expression of the form [178]:

$$v_h = v_0(1 - \varphi_v)^\sigma \tag{38}$$

Table 1.

System	Advantages	Disadvantages
Tubular Bowl	a) High centrifugal force b) Good dewatering c) Easy to clean d) Simple dismantling of bowl	a) Limited solids capacity b) Foaming unless special skimming or centripetal pump used c) Recovery of solids difficult

Chamber bowl	a) Clarification efficiency
	  remains constant until
	  sludge space full
	b) Large solids holding capacity
	c) Good dewatering
	d) Bowl cooling possible
		a) No solids discharge
		b) Cleaning more difficult than tubular
		  bowl
		c) Solids recovery difficult

Disc-centrifuges	a) Solids discharge possible
	b) Liquid discharge under
	  pressure eliminates
	  foaming
	c) Bowl cooling possible
		a) Poor dewatering
		b) Difficult to clean

Scroll discharge	a) Continuous solids discharge b) High feed solids concentration	a) Low centrifugal force b) Turbulence created by scroll

where v_h is the hindered sedimentation velocity and σ is a geometric factor. For monosized, rigid, spherical particles, $\sigma \simeq 4.6$ whereas for non-rigid, non-spherical particles, σ could vary from about 10 to 100 resulting in large errors where Stokes law is assumed. Using Stokes law as a basis for machine design it can be seen that improved recovery can be achieved by either reducing the settling distance or increasing the sedimentation velocity. The main machine configurations are shown in Fig. 26 with some of the advantages and disadvantages shown in Table 1. A description of their operation is given by Wang et al. [144]. A reduction in settling distance is achieved by flowing in thin channels such as in the disc machines (c) and (d), although this can result in deviations from Stokes law due to particle-wall interactions. Increasing the sedimentation velocity can be achieved by either increasing the particle diameter or density, reducing the fluid viscosity, or by increasing the centrifugal force. This last method can be carried out by increasing the speed of rotation or the radius of settling, but is limited by the materials of construction and the requirements of mechanical strength, chemical resistance and sanitary design.

Relating the description of sedimentation by Stokes law to centrifuge operation and scale-up is typically carried out using the Σ-concept [174, 179, 180], although a grade

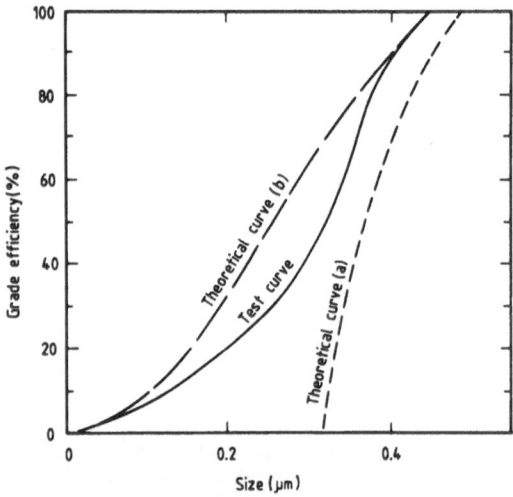

Fig. 29. Comparison of experimental test and theoretical grade efficiency curves for a tubular bowl centrifuge [181].

Curve a) using parabolic velocity profile:

$$G(d) = \frac{r_3^2 - 1.05 \times 10^{-4} \left(\dfrac{18\mu Q}{d^2 \Delta\varrho\, \omega^2} \right)^{5/3}}{r_3^3 - r_1^3} \; ;$$

Curve b) using plug flow velocity profile:

$$G(d) = \frac{r_3^2}{r_3^2 - r_1^2} \left[1 - \exp\left(\frac{-2 \Delta\varrho\, \omega^2}{18\mu} K d^2 \right) \right]$$

where K, volumetric ratio of effective bowl capacity to flow rate; r_3, bowl wall radius; r_1, fluid surface radius efficiency method can also be used [181]. The grade efficiency method accounts for the existence of a particle size distribution and the probability nature of particle capture. The gravimetric grade efficiency function G(d) is the curve of mass efficiency for each particle size, d_i:

$$G(d)_i = (mass\,separated)_i/(mass\,in\,feed)_i \tag{39}$$

Functions can be derived for any given geometry by the application of Stokes law and different liquid velocity profiles. Two such functions are shown as curves a and b in Fig. 29. As for the Σ-concept, this method is limited by the validity of the assumptions involved. Alternatively experimental functions can be generated from feed and overflow particle size distributions, Fig. 29 (test curve), and used to predict performance for varying feed distributions and feed rates [182]. A similar approach is used by Sokolov and Borisenko [183], incorporating a particle size distribution function with the axial hydrodynamic flow in a graphical analysis for tubular bowl centrifuges. Use of grade efficiency plots and three dimensionless parameters for the scale-up of geometrically similar disc-stack centrifuges has been recently proposed by Gupta [184].

5.3 Fluid Dynamics within Centrifuges

Prediction of the separation performance of centrifuges is based on semi-empirical relationships derived from simplified theoretical descriptions of the fluid flow. However, the understanding of these characteristics is continually improving due to the development of theoretical models which use a Navier-Stokes description of the hydrodynamics and account for the influence of specific machine parameters. An understanding of the mathematical descriptions is not necessary for the design of process systems but knowledge of the concepts involved and their significance in limiting machine performance is of benefit in the selection of equipment to handle specific precipitates.

5.3.1 Tubular Bowl Centrifuges

The tubular bowl centrifuge, as shown in Fig. 26 (a), has a relatively simple configuration with the suspension being pumped to the rotor via a nozzle through the bottom and solids are sedimented to the rotor wall as the liquor flows upwards. Simple descriptions based on annular plug flow or thin layer models [176] are inadequate due to end effects, the swirling and velocity lag of the liquid [185], and the effect of solids build-up in the bowl. Experiments by Sokolov et al. [186] showed that the simple flow models are invalid and in particular that the surface flow regime is related to the feed arrangement. Golovko [187] found that the use of a cross-shaped insert to disperse the feed and an inlet device to accelerate the liquid and so reduce lag, resulted in improved efficiency as defined by the increase in the Σ-value, Fig. 30. The disadvantage of these feed devices for protein precipitate aggregates is that the degree of local turbulence is usually sufficient to cause break-up resulting in some cases in a reduced efficiency. Full length winged inserts are also used to minimize fluid lag effects.

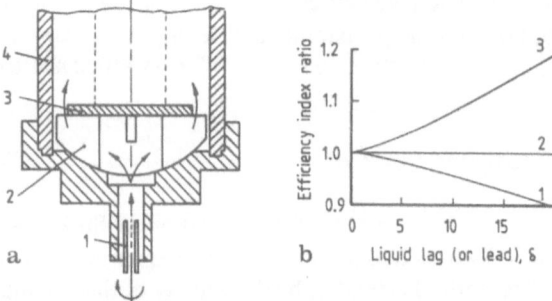

Fig. 30 a. Lower part of the rotor of a tubular bowl centrifuge with inlet device: 1) feed nozzle, 2) vane, 3) disc, 4) rotor. **b** Dependence of efficiency index ratios on the maximum lag or lead of liquid: 1) smooth rotor, 2) vaned rotor, 3) rotor with inlet device as in **a** [187]

Assuming a plug flow liquid profile the separation limit, d_{crit}, of a tubular bowl centrifuge can be given by [181]:

$$d_{crit} = \left[\frac{18Q\mu}{\pi \, \Delta\varrho \, \omega^2 l} \frac{\ln (r_3/r_1)}{(r_3^2 - r_1^2)} \right]^{1/2} \tag{40}$$

where Q, throughput; r_3, tube wall radius; r_1, liquid overflow radius; l, bowl length. The limit of separation is the maximum particle size that would escape with the overflow.

5.3.2 Multichamber Centrifuges

Descriptions of the hydrodynamic characteristics of multichamber centrifuges, Fig. 26 (b), can be simply considered as an extension of those for the tubular bowl system. However this gives a rough approximation only. Multichamber bowls operate in a flooded condition and so there is no air-liquid interface as in the tubular bowl machine. The different axial velocity profiles and the changing radial distance and solids concentration in moving from the inner to outer chambers make any analysis difficult. Comparisons of the relative efficiencies of tubular and multi-chamber centrifuges by assuming given hydrodynamic profiles can only give an indication of possible machine performances since the separation efficiency of the high speed, small diameter tubular bowl centrifuge is critically dependent on selecting an approximate sedimentation radius. Use of the liquid overflow diameter (i.e. thin-layer model) may predict a lower performance than is actually achieved, since the presence of an inlet device as shown in Fig. 30 can be considered in simple terms to increase the effective sedimentation radius.

5.3.3 Disc Centrifuges

Disc type centrifuge designs, Fig. 26c and d, represent a considerable degree of complexity with the central feature being a set of conical discs used to both reduce the sedimentation distance and increase the settling area. Several workers have recently calculated the velocity profiles between the discs accounting for the Coriolis force and shear force effects [188, 189, 190, 191]. The Coriolis force results in a deformation of the velocity profile which depends on a dimensionless hydrodynamic parameter:

$$\lambda = h \sqrt{\frac{\omega \, \mathrm{Sin} \, \theta}{v}} \qquad (41)$$

where h, disc separation; v, kinematic viscosity; ω, angular velocity; θ, included angle between axis and disc surface. The velocity profiles for two values of λ are shown in Fig. 31a, b [191]. The occurrence of radial backflow and the relatively high value of the circumferential velocity for high values of λ due to the increasing Coriolis force have a significant effect on separation performance. Industrial centrifuges operate in the range of 6 to 15 for λ with average radial velocities from about 0.05 to 0.3 m s^{-1} and so the transition from laminar to turbulent flow may be critically influenced by the circumferential velocity of fluid flow [191]. The advantage of having spacers in the gap between the discs is that they suppress this circumferential velocity component. The parameter, λ, shows the influence of disc spacing which is found in practice but which is not accounted for in the Σ-type analysis.

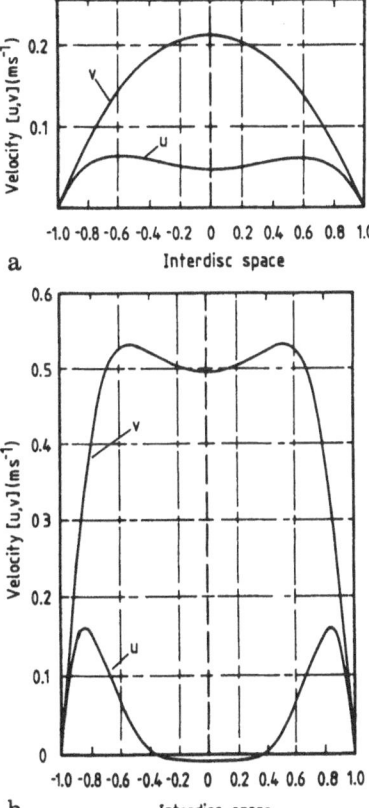

Fig. 31a and b. Velocity profiles in radial and circumferential directions for the flow between centrifuge discs. a $\lambda = 4$, $\bar{u}_a = 0.05$ m s^{-1}; b $\lambda = 10$, $\bar{u}_a = 0.05$ m s^{-1}; u, radial velocity; v, circumferential velocity [191]

Another significant effect is that of the entrainment of particles into the discharging stream, lowering the separation efficiency [190, 192, 193]. Brunner and Molerus [191] showed that because of flow instabilities in the gap, vortices are formed. These vortices transport fine particles (<1 μm) to the disc surface where they settle out, and so assist in separation efficiency. However, coarse particles (>1 μm) are preferentially remixed by the vortex resulting in a reduction in separation efficiency. This effect resulted in at best a 90% degree of separation at the separation limit, as defined by Eq. (42). Carlsson [190] has developed a model to account for both the negative effect of the shear forces at the disc surface and the influence of separation effects outside the disc stack. His model predicts that separation outside the disc stack results in identically transformed particle size distributions entering the different disc channels assuming that the same flow rate is obtained in every inter-disc channel. On the other hand Skvortsov [194] analysed the motion of the solid component in the sludge space and showed that there is a non-uniform distribution of the solid product entering the disc stack. Brunner and Molerus [191] have indicated that spacers which divide the discs have a beneficial effect on separation, and this approach has been developed as the concept of corrugated disc stacks as reported by Vinogradova et al. [195]. They found that improved efficiency was achieved when the inter-disc space consisted of a system of conical radial channels and suggested that the improvement was due to minimizing the interactions between the dispersion and sedimented sludge, and the effectively zero tangential velocity of the fluid with respect to the disc surface. Other workers [196] have accounted for the loss of efficiency due to machine vibrations and the effect of suspension solids volume on the flow distribution [197].

If plug flow is assumed between the discs then the separation limit for a disc centrifuge can be given by [191]:

$$d_{crit} = \left[\frac{27 Q\mu \tan \theta}{n\pi (r_2^3 - r_1^3) \Delta\varrho \, \omega^2} \right]^{1/2} \tag{42}$$

where n, number of disc channels; r_2 disc outer radius; θ, included angle between axis and disc surface.

5.3.4 Scroll Discharge Centrifuges

In the horizontal, screw-conveyor type, solid bowl centrifuges the feed enters a central inlet pipe, is accelerated to bowl speed through a feed chamber and the particles sediment to the bowl wall. The sludge is continuously removed from the bowl by a screw-conveyer which rotates at a slightly faster or slower speed than that of the bowl. Liquid discharges freely or via a centripetal pump. These centrifuges can be considered fundamentally as tubular bowl machines with special solids discharge features. However, this is a crude approximation since turbulence is created by the relative motion of the conveyer, the conveyer flights take up space in the bowl, the spiral liquid path is difficult to assess and hindered settling occurs due to the usually high feed solids concentration. Although these machines have a wide use in the food and water-treatment industries few theoretical descriptions have been developed. Sokolov and Semenov [198] have derived a function for the efficiency, accounting for the action of the rotating screw on the hydrodynamics and allowing for a feed particle size distribution. This analysis assumes that recovery occurs in the cylindrical section only. Semenov and Karamzin [199] accounted for the full rotor length including the conical section in their equation for the critical particle diameter and purification coefficient. Other workers [200] have studied the relation between the hydrodynamics in the bowl and the flocculation of the solid phase. They calculated the distribution of velocity gradients in a scroll centrifuge using a simplified analysis which did not include the effect of the scroll but accounted for the existence of fluid lag. Figure 32 shows that the calculated distribution of velocity gradients has a maximum away from the free surface due to the ratio between the velocity gradient components in the radial and longitudinal directions. The gradient becomes less sharp and the maximum moves towards the wall in moving away from the conical section. The relatively low velocity gradient values calculated in this analysis support the use of flocculants to improve the sedimentation within the bowl [201] as widely used in the water treatment industry and strengthen the view that shear disruption occurs in the feed zone. Further evidence of the effect of shear disruption on machine performance is given by Gösele [202] and Bell and Dunnill [124]. Gösele also indicates the importance of turbulence on scale-up of decanter centrifuges and suggests the use of the power dissipated per unit volume as a scale-up parameter.

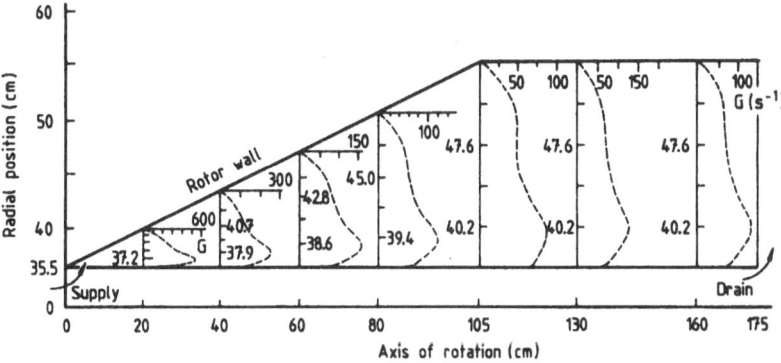

Fig. 32. The velocity gradient distribution at locations along the axis of the rotor of a centrifuge [200]. Centrifuge: NOGsh-1100A; speed, 750 rpm; feed rate, 50 m³ h⁻¹; G, velocity gradient (s⁻¹); – – – – velocity gradient profile

Since decanter centrifuges are normally discharging concentrated sludges, an understanding of the dewatering mechanisms for different types of sludges is important. Redeker [203] indicates that dewatering involves the flow of liquid along the conveyor flights in the opposite direction to the solids transport. Another mechanism involves drainage through the gap between the scroll and the bowl wall [204]. This mechanism would be more suited to non-compressible coarse particles than to the typically fine compressible protein sludges. Gösele [202] considered the importance of both compressive and drainage mechanisms in relation to machine scale-up.

5.4 Modifications of Industrial Centrifuges

The specific nature of protein precipitates as considered in Sect. 5.1 has required the development of many modifications to the standard models. Some of these modifications are considered here.

5.4.1 Tubular Bowl Centrifuges

The principal objection to the operation of tubular bowl centrifuges is the batch mode of operation whereby the machine must be stopped and dismantled to effect the solids discharge, resulting in both operational hazards and production downtime. Several alternative methods of operation have been studied. Vallet [205] developed a method of solids resuspension by agitation with a solution being circulated through the bowl. However this system requires the removal of the bowl from the drive system and also removal of the three-wing insert assembly from within the bowl. Miller et al. [206] studied the use of a flushing fluid for aseptic operation or a piston discharge for non-aseptic operation. The piston system involves a manual withdrawal of the solids whereas the flushing fluid operates in a closed circulation system. Both methods require the bowl to be removed from the drive assembly. The flushing method also suffers from the effect of foaming associated with the use of spray nozzles in protein systems. Hoare [207] investigated the use of folding vanes to discharge the solids. Most standard tubular bowl centrifuges also suffer from the effect of high turbulence in the feed zone and are operated on the assumption that centrifugal force is available to recover disrupted particles. However, any disruption must result in a reduction in the separation efficiency. The use of a conical feed section similar to those designed for the nuclear industry [208] may provide a solution to this problem, by a more gentle acceleration of the feed to bowl speed.

5.4.2 Vortex Clarifiers

A tubular bowl vortex clarifier may offer a suitable alternative for applications involving proteins which are sensitive to denaturation by air-liquid interfaces and require aseptic discharge conditions. A machine is already in use in the water treatment field [209], and with appropriate modifications may be suitable for the separation of some protein precipitates. The main limitation of these machines is the low centrifugal force.

The hydrocyclone [210] is another device which utilizes vortex flow to separate solids from liquid streams and which may offer suitable separation or pre-concentration of solids in some protein systems. It is unsuitable for low density, submicron particles and its stage efficiency is lower than that for centrifuges. However this can be overcome by multistaging systems. Its main use to date outside the mineral industries is in the starch industry.

5.4.3 Multichamber Centrifuges

The multichamber centrifuge was developed to give similar clarification efficiency to the tubular bowl centrifuge but with a much larger sediment holding space. The high efficiency is achieved by long residence times which often result in unacceptable temperature rises. This has been overcome by direct cooling of the bowl, centripetal pump and the frame [211]. The problems of complete drainage of the bowl fluid and removal of the sedimented solids are the main operational difficulties with these machines. In some models the liquid in the bowl at shutdown drops from a hole near the central axis as the bowl slows down. In others it must be siphoned off by removing the machine hood and inserting a tube into the bowl. Both methods are unsuitable for wet sludges which tend to dislodge from the walls with any agitation and so become redispersed with the residual fluid. The use of liners to improve sludge removal has not found general application due to the problem of their retention. The application of hermetic seals in both multichamber and disc centrifuges has minimized air entrainment and maintained air-tight operation. This reduces protein denaturation and results in significantly less turbulence in the machine feed zone, but at a higher cost.

5.4.4 Disc Centrifuges

Some problems of batch operation were overcome with the introduction of nozzle and intermittent discharging disc centrifuges. As with the multichamber centrifuge the requirement of low throughput for difficult solids results in significant fluid temperature rises through these machines and has necessitated the introduction of cooling of the bowl, centripetal pump, hood and sediment catcher [212]. The cooling of the sediment catcher is necessary for "sticky" protein solids where "burn-on" can occur. The low throughputs have been due to the relatively low "g" force of these machines. Recent advances with the introduction of new grades of high strength stainless steel and changes in the configuration of the discharge mechanism have improved the clarification efficiencies by increasing the relative centrifugal force from about 10,000 to 14–15,000 g and enabling discharge time control to 0.1 s [213, 214, 215]. The discharge of concentrated protein sludge also creates problems with the operation of intermittent discharging centrifuges and requires wide discharge ports, accurately controlled partial desludging and specially designed discharge chutes to maintain sludge flow. These improvements are found on the larger machines only. Small machines suffer from difficulty with the rapid operation of the bowl discharge mechanism required for partial desludging and so usually produce wetter sludges. The desludging actuators use hydraulic, pneumatic or spring systems. Water contamination can be a problem with hydraulic systems, although the introduction of special packing seals has enabled the use of alcohols as operating liquid in some machines. The control of desludging is carried out using an automatic timer unit, a turbidity meter on the clarified liquor, or an internal sludge sensing probe. The turbidity method suffers from the disadvantage of operating as a feedback control system and is activated only after a product deterioration has occurred. The automatic timing method relies on constant process conditions of feed rate, solids concentration, particle sedimentation and sludge compressibility characteristics. The internal probe senses the level of solids in the bowl and so offers better control in systems where the feed solids content varies. Most of the disc centrifuges are capable of cleaning in place utilizing detergents, alkali, acids, or proteolytic enzymes, depending on the type of process. Cleaning of intermittent discharge machines is usually achieved by operating through a sequence of desludging cycles with cleaning solutions in the bowl. Some machines can be flooded in the shut-down condition to enable access of the cleaning fluid to all areas. Cleaning in some of the larger nozzle discharge machines can be achieved by reducing the bowl speed and reversing the flow direction back through the nozzles. Steam sterilization to 120 °C is also possible in machines with the appropriate seals. This operation is carried out in the shut-down mode. A significant problem with intermittent disc centrifuges is the collection of bioactive sediments which present a hazard. Sealed, sterile systems with sediment collection chambers designed to dissipate the aerosol wave generated during discharge, are available from some companies.

The influence of particle disruption due to the shear field in the machine feed zones can result in a significant loss in efficiency. Some reduction in shear is achieved by the use of hermetic feed systems and further improvements may be obtained with disc machines by the incorporation of a conical shaped inlet device which accelerates the feed to bowl speed, Fig. 33a [216].

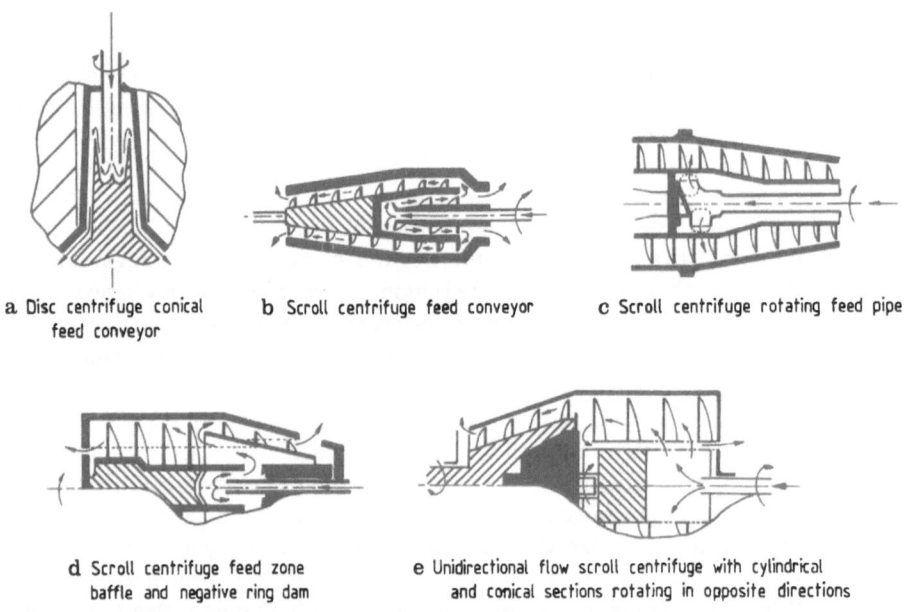

a Disc centrifuge conical
  feed conveyor

b Scroll centrifuge feed conveyor

c Scroll centrifuge rotating feed pipe

d Scroll centrifuge feed zone
  baffle and negative ring dam

e Unidirectional flow scroll centrifuge with cylindrical
  and conical sections rotating in opposite directions

Fig. 33a—e. Centrifuge feed geometries designed to reduce the effects of shear

5.4.5 Scroll Discharge Centrifuges

The influence of turbulent shear in the feed zone of scroll discharge centrifuges can produce significant particle disruption [124, 202], and in some systems it can be reduced by the use of a feed conveyor, Fig. 33b. Other methods of reducing shear disruption involve the use of a rotating feed pipe [201] Fig. 33c, the insertion of a baffle in the bowl at the feed zone to isolate the effects of feed entry from the sedimented sludge [217] Fig. 33d, and the use of unidirectional flow centrifuges with axial discharge tubes [218] Fig. 33e. In some systems these axial tubes may result in solids blockage. The entry zone baffle, Fig. 33d may also be used as a negative ring dam, in the recovery of fine light solids. In this configuration the solids discharge port is placed at a greater radial distance from the bowl axis than the liquid discharge port. This results in improved recovery for some solids such as protein precipitates which exhibit difficult rheological characteristics. It also enables the use of a deeper pond depth. Pressure discharge machines are also available where system characteristics such as foaming are present, as well as gas tight models for explosion proof operation.

5.5 Zonal Ultracentrifuges

A system which has not been considered in the previous discussion is the zonal ultracentrifuge. These machines can develop very high relative centrifugal forces (up to 90,000 g) and offer a method of recovering very fine, light, particles. A review of the types of equipment available and hydrodynamic considerations has been given by Hsu [219]. Charm and Matteo [220] have reported the recovery of an ammonium sulphate precipitated protein at $10-20 \, l \, h^{-1}$ with a small, 1 l liquid bowl capacity. Cline [221] also reported the recovery of a barium precipitated virus preparation at $5 \, l \, h^{-1}$. These machines operate in a flooded condition and so do not suffer from the same foaming problems associated with many tubular bowl centrifuges.

6 Processes Involving Protein Precipitation

The key role of precipitation in protein processing may be illustrated by several examples. The processes were developed prior to the availability of any detailed design data and were evolved by a largely empirical approach based on experience. Nevertheless they indicate the type of situations in which a biochemical engineering approach can contribute.

6.1 Production of Soya Protein Isolate

The preparation of vegetable proteins free of carbohydrate and other constituents of the original sources has become increasingly important recently. The production of soya protein isolates is now established industrially and though the details of commercial operation are not released the broad outlines are well known.

Soya beans are first cracked, dehulled, flaked and then oil is extracted with solvents such as hexane. Done under low moisture conditions this does not lead to serious protein denaturation though the removal of solvent can entail some damage. The defatted meal contains 50% protein, a maximum of 1% oil and the remainder is made up of soluble and insoluble carbohydrates (30%), water and minor components. The insoluble carbohydrate is removed by dissolving the protein and soluble sugars in an aqueous alkali solution at pH 7 to 9 and removing the insoluble carbohydrate using scroll or intermittent discharge disc centrifuges. The protein is then recovered by isoelectric precipitation at pH 4.6–4.8 to give a product of 90–95% protein but with a yield of only 30–40%.

In most industrial processes the precipitation step appears to be batchwise with the continuously operating centrifuges servicing several tanks working out of phase. A precipitate slurry of about 1% dry solids may be recovered in the intermittent discharge of the disc centrifuge as a sediment of nearly 20% dry solids with the machine operating at an underflow to overflow ratio of 1 to 11 [222].

Recent work has shown the importance of choosing the acid carefully [20] to avoid damage due to particular anions. Information is now available on the kinetics of soya protein precipitation [99] (Sect. 3.3), the variation of density with particle size [114] (Sect. 5.1), the effect of shear on soya protein precipitates [89, 92] (Sects. 3.2 and 4.1) and the relationship of shear effects and precipitate sludge rheology to centrifugal recovery in disc and scroll centrifuges [124] (Sect. 5.1). The availability of such biochemical engineering data would make the operation of a continuous precipitation process easier and with it the ability to automate and closely control a large scale industrial plant.

6.2 Continuous Fractionation of Human Blood Plasma

Scale-up of precipitation of proteins by batchwise addition of reagents is not straightforward in ethanol fractional precipitation of human plasma proteins where the reagent can cause damage. The necessary reduction of temperature to subzero is time consuming with batch stirred tanks of 2–6 m³ which are commonly used. The degree of agitation to avoid local heating can lead to rapid replenishment of the upper surface causing protein denaturation and even foaming with severe consequences. With less agitation local heating due to ethanol-water interaction will be significant and pH and temperature control poor. With these problems in mind Watt and co-workers [223] have developed a continuous fractional precipitation process which consists of a number of unit operations imposed on the flowing stream of plasma (Fig. 34). These include: (a) flow metering of ethanol, plasma and reagents (b) pH adjustment (c) continuous mixing and cooling (d) continuous-flow ageing.

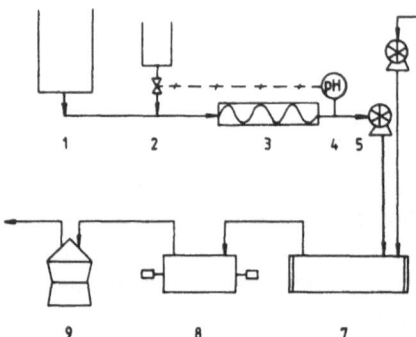

Fig. 34. Continuous flow process for the fractionation of human plasma proteins: (1) plasma feed tank; (2) titrant feed tank; (3) static mixer; (4) pH controller; (5) metering gear pump for plasma; (6) metering gear pump for ethanol; (7) mixer-cooler; (8) acoustic vibrator; (9) multichamber centrifuge [223]

The plasma is mixed with acids and buffers using an in-line static mixer sized to give complete mixing with a dead-time of about 10 s. The response of the downstream pH electrode is dependent on the ethanol and protein concentrations and, if precipitation has occurred, by its extent. Gear pumps are used to meter the plasma and ethanol streams to a second static mixer of a different design in which the mixing elements are integrated into a stainless steel shell and tube heat exchanger. The precision of temperature control is determined by the response time of the temperature sensor and the sensitivity of the refrigerant control valve and actuator. An acoustic conditioning cell [131] is used to simulate the effect of long term ageing prior to solid-liquid separation in multi-chamber centrifuges incorporating direct refrigeration of the bowl. For ease of working all the elements of the precipitation system shown in Fig. 34 are mounted within a mobile unit and several units may be linked to accomplish multi-stage fractionation of plasma. Typical single stream flow rates are $30 \, l \, h^{-1}$ and all the sensing elements are linked to a process control computer which calculates the necessary actions and drives the control valves [223].

There is little doubt that continuous precipitation represents a better process engineering approach to human plasma fractionation than batch operation. Reservations have to do with the special nature of the product and possibly with changes in the nature of the precipitates that affect recovery. Individual protein fractions other than albumin contain a number of proteins many of which are poorly defined. Their therapeutic value can only be judged clinically and any alteration to a long established procedure raises the possibility of changing this clinical effect. There is no evidence to date of this being the case. As indicated in Sect. 4.1.2 the nature of the precipitate particles formed during continuous isoelectric precipitation of soya in a tubular reactor is less compact than that formed in a batch stirred tank and more susceptible to shear break-up. It is not known whether this behaviour is exhibited by ethanol precipitated plasma proteins or whether substitution of a continuous stirred tank reactor would produce a precipitate more similar to batch prepared product.

6.3 Continuous-flow Isolation of Intracellular Enzymes

It is a common experience that when a laboratory scale isolation scheme for an intracellular enzyme is scaled up the yield and specific activity suffer [224]. Part of this loss is evidently due to the greater time taken to process large quantities batchwise even with the use of industrial process equipment. Adoption of continuous-flow

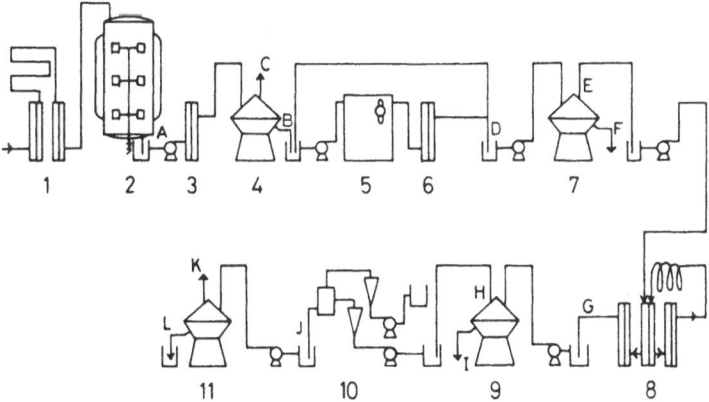

Fig. 35. Flow diagram for the continuous isolation of β-galactosidase from *E. coli*. (1) continuous-flow plate media sterilizer; (2) 1000 l fermenter; (3, 6) heat exchangers; (4, 7, 9) disc-bowl centrifuges; (5) homogenizer; (8) heat exchanger and holding coil; (10) continuous-flow mixer; (11) disc-bowl centrifuge or tubular centrifuge [154]

processing can reduce the time taken to below that achieved in the laboratory, it allows the use of smaller equipment for a given scale and can give a more reproducible product.

Continuous precipitation has been used as one stage in the continuous production and isolation of intracellular *Escherichia coli* β-galactosidase [154]. The process based on a 1000 l fermenter operated as a chemostat is illustrated in Fig. 35. The protein was precipitated in a CSTR with a helical ribbon stirrer by saturated ammonium sulphate following heat treatment to degrade viscous nucleic acid. The heat treatment shifted the position of salting-out substantially compared to that observed with unheated material. Centrifugal recovery of the precipitate was by intermittent discharge disc centrifuge operating at a low flowrate. With partial discharge of solids a ratio of overflow to discharge solids of 14:1 was observed. With fully continuous operation the time taken to produce a precipitated enzyme following harvesting, cell rupture, debris removal and heat treatment was 2 h.

A continuous process involving two stage fractional isoelectric precipitation has been used to purify a prolyl-tRNA synthetase enzyme from the bean *Phaseolus aureus* [225]. The flowing acid and protein streams were mixed by impingement from opposing jets and the precipitate separated in intermittent discharge disc centrifuges after a brief holding time in short coils. The lower process time (3 h) compared to a batch process (9.5 h) raised the specific activity of the labile enzyme sixteen-fold and the yield 50–60 times.

Continuous enzyme processing evidently can improve the efficiency of intracellular enzyme purification. However it is less flexible than a batch plant and greater plant reliability is required so that it is best justified when a given intracellular enzyme is to be manufactured on an industrial scale over a period of years. It is consistent with this view that continuous fermentation and isolation is used for the commercial production of glucose isomerase [226].

7 Conclusions

7.1 An Approach to Process Design

Efficient process operation requires the optimum design of systems for the precipitation and recovery of protein precipitates. This involves a careful consideration of the precipitation method, the characteristics of the protein precipitate and its interaction with the chosen centrifuge.

7.1.1 Precipitation Methods

The physico-chemical aspects of protein precipitation are represented in a simplified general way by Fig. 36. The reduction in protein solubility on the addition of the major classes of precipitants is approximately according to a power law relationship of the general form:

$$\log S/S_0 = KR \tag{43}$$

where R is the reagent concentration and K is a constant. The term S_0 is the hypothetical solubility of the protein at zero reagent concentration. The value of K for a given

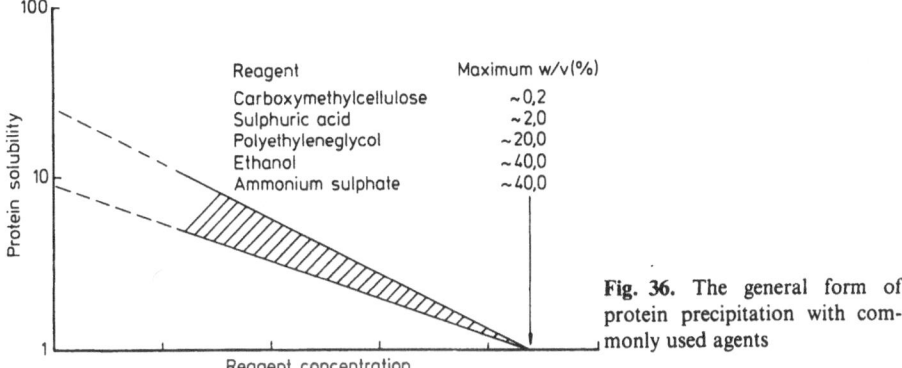

Fig. 36. The general form of protein precipitation with commonly used agents

reagent will be influenced by other parameters of the physico-chemical environment and as shown in Sect. 4.1.1 even the position of the precipitation curve may be shifted by the contacting conditions.

The concentrations required to produce total precipitation are generally of the magnitude shown in Fig. 36 for representative reagents. The reagent costs increase in the order sulphuric acid, ammonium sulphate, polyethylene glycol, ethanol, carboxymethylcellulose, so that the cheapest method of precipitation will be by acid reduction of pH to the isoelectric point followed by precipitation using ammonium sulphate, carboxymethylcellulose, polyethylene glycol and ethanol. In practice as indicated in Sect. 2 many considerations determine the choice of reagent. Establishing the specific form of Fig. 36 in a particular case of interest under the environmental conditions to be used finally on a large scale is an essential first step in process design. It must be done for each of the potentially useful precipitation reagents and for any variants of the physico-chemical environment which may alter the precipitation curve. Following the establishment of the precipitation curve, the kinetics of precipitation should be evaluated at the position on the curve for which the process is to be designed.

The method of protein precipitation in large scale processes is usually inherited from the development studies carried out at the laboratory bench. In many cases this takes no account of the special problems certain reagents create at the industrial scale such as flammability, high viscosity, high liquid density and corrosiveness (Sect. 2.2), and the precipitation yield obtained at the laboratory scale may not be achieved due to particle recovery limitations. For this reason the precipitation method selected may have to be a compromise between precipitate recovery efficiency and precipitation yield. An assessment of the precipitate sedimentation characteristics should preferably be made alongside the selection of the precipitation method. This can be readily performed by simple comparative tests in a laboratory bench centrifuge.

7.1.2 Reactor Design for Precipitate Formation, Ageing and Recovery

As discussed in Sect. 4 the required mixing conditions during the nucleation stage will be determined by several factors such as the kinetics of nucleation, the importance of shear or thermal denaturation and the method of reagent addition. The nucleation kinetics and denaturation effects can be assessed at the laboratory bench. The effect of contacting procedure and mixing conditions may be significantly affected by scale-up.

Selection of a continuous or batch process will depend on the reaction kinetics and on the particular system requirements, i.e. throughput, single or multi-product capability, the desirability of automation, etc. With rapid precipitation reactions the use of continuous tubular reactors offers a system with low hold-up and so facilitates control of process variables. These reactors are also amenable to scale-up either by multiple parallel units or by an increase in diameter using the average shear rate and residence time as a scale-up parameter for ageing processes. Defined mixing conditions in tubular systems can be achieved by the use of turbulence promoters with the subsequent particle growth to a maximum size being controlled by ageing.

In general the following conditions should be considered in order to prepare an aggregate of maximum size and strength:

a) the precipitating reagents should be added under mixing conditions that minimize denaturation yet ensure good mixing.
b) the shear rate should be reduced as soon as possible after nucleation to promote orthokinetic growth in a controlled manner.
c) the aggregated precipitate should be aged for sufficient time to maximize its strength.
d) exposure to all high rates of shear should be minimized by careful design and operation of the suspension piping system and the centrifuge feed zone.

For a given protein solution and method of precipitation the particular reactor configuration and extent of ageing will result in a particle size frequency distribution, $q_0(G, t)$ which is a function of the rate of shear, G, and the time of exposure to shear, t. The shape of this distribution will be defined by the particular conditions and time of ageing, with long periods of ageing usually resulting in a reduction in the number of fine particles, given that significant shear break-up is not occurring concurrently.

Design of a process to prepare and recover the precipitate can be made by defining a limiting characteristic for centrifuge performance and using this to establish a minimum particle size (or settling velocity) for the particle preparation. The critical diameter (Sect. 5.3), which corresponds to the smallest particle size that is just recovered can be used to define centrifuge performance. The ageing process would then be designed to ensure that say 95 % or 99 % of the particles were larger than the centrifuge critical diameter. This would require information about the growth kinetics of the protein precipitate. Other factors which must be accounted for include the existence of aggregate strengthening due to ageing under defined shear conditions and the possibility of aggregate break-up. If break-up does occur in feed piping or in the high shear field of the centrifuge feed zone, then this must be accounted for in designing the aggregate growth and ageing processes. Ideally a knowledge of break-up mechanisms is necessary to predict whether fine particles or large fragments will be produced by break-up. This information, along with the growth kinetics data would enable reactors to be designed using the concepts discussed in Sects. 3 and 4. Since particle size analysis is not practically possible for all protein systems a similar design basis could be established using sedimentation velocity as the limiting characteristic.

For a detailed process design the centrifuge critical diameter should be replaced by the centrifuge gravimetric efficiency curve (Sect. 5). This describes the separation performance for the complete range of particle sizes and so would enable a more accurate estimate to be made of the total separation efficiency for a given particle size distribution.

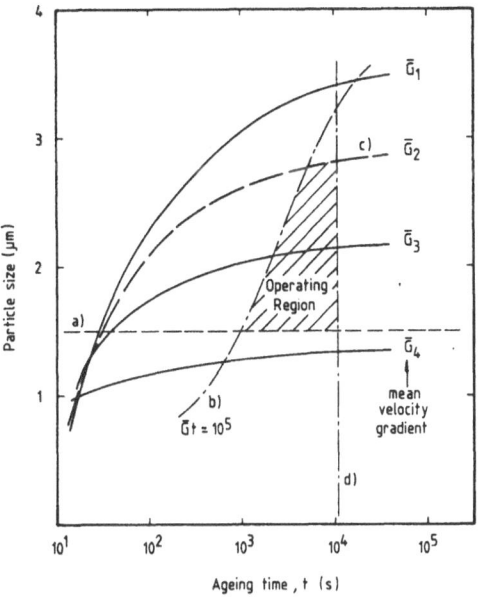

Fig. 37. Precipitate preparation process design. Particle size versus ageing time for batch precipitation. Operating region defines the conditions of ageing necessary to prepare a recoverable precipitate for a given centrifuge. Limits: a) minimum size for centrifugal recovery, b) optimum strength parameter ($\bar{G}t \simeq 10^5$), c) minimum stirrer speed for efficient mixing, d) maximum ageing time limit

If all of the above relationships can be established then the operating region for the ageing process prior to separation in a given centrifuge can be shown graphically as in Fig. 37. This figure shows a typical relationship between particle size, and ageing time for different mean velocity gradients in the ageing vessel. Four particle growth curves are indicated at mean velocity gradients $\bar{G}_1$, $\bar{G}_2$, $\bar{G}_3$ and $\bar{G}_4$. The particle size increases at a given mean velocity gradient for increasing ageing time. However, an increase in the mean velocity gradient from $\bar{G}_1$ to $\bar{G}_4$ results in a reduction in the final particle size. Line (a) is the minimum particle size for centrifugal separation and corresponds to the critical particle diameter. Line (b) is the curve for $\bar{G}t \simeq 10^5$, that is the minimum value of the product of the mean velocity gradient in the vessel and the ageing time, required to optimize the aggregate strength. Line (c) can be defined for processes where reagent mixing and ageing occur in the same vessel. This limit corresponds to the minimum mean velocity gradient required to achieve good mixing and avoid denaturation effects due to high localized temperature and reagent concentrations. Line (d) is the maximum time limit for the ageing process and will be defined by the particular process operating requirements. These lines define the boundaries for the design of the ageing process.

A theoretical approach to process design can be made by considering the change in the particle size distribution for each stage of preparation and recovery. Use of Eqs. (27) and (28) from Sect. 3 enables the particle size distribution $q_0(G, t)$ to be defined for a given ageing process. The characteristics of the feed system to the centrifuge and the centrifuge feed zone may result in aggregate break-up, leading to a change in this feed distribution as defined by some break-up transfer function, $g_B(v)$:

$$q_0(G, t) \rightarrow \boxed{g_B(v)} \rightarrow q_1(v) \qquad (44)$$

where v is the sedimentation velocity of the particles, which accounts for the existence of both size and density distributions. The resulting particle size distribution is then subjected to the particular sedimentation characteristics of the given centrifuge. This results in a further change in the particle size distribution, defined by a centrifuge clarification transfer function, $g_C(v)$

$$q_1(v) \rightarrow \boxed{g_C(v)} \rightarrow q_2(v) \qquad (45)$$

If both transfer functions, $g_B(v)$ and $g_C(v)$, are known, then the necessary initial centrifuge feed distribution, $q_0(G, t)$, can be defined to achieve a given recovery. A description of $g_C(v)$ for disc centrifuges has been given by Murkes and Carlsson [227]. However a knowledge of the ageing process to obtain $q_0(G, t)$ requires better understanding of the growth and break-up processes involved for protein precipitates.

The design of protein precipitation and precipitate recovery processes requires information from physical biochemistry, molecular biophysics, colloid chemistry and biochemical engineering. Some of the key mechanisms must still be inferred from more detailed studies of other biological systems or even non-biological ones. Nevertheless a scientific framework is beginning to emerge for protein precipitation. It will increasingly allow laboratory tests to be translated into efficient engineered processes.

8 Acknowledgements

The authors are grateful to Professor J. W. Mullin, Dr. J. Gregory, Dr. P. R. Foster, Dr. F. Rothstein and Dr. K.-H. Brunner for helpful comments and to Mrs. C. J. Henderson for the manuscript preparation of this review.

9 Nomenclature

C   concentration, $kg\ m^{-3}$

d, d_x particle diameter, where x refers to the wt % oversize value, m

D_s dielectric constant

D   diffusivity, $m^2\ s^{-1}$

f_{ij} interaction coefficient, Eq. (5)

$\bar{G}$ mean velocity gradient, s^{-1}

G   shear rate, s^{-1}

G(d)   gravimetric grade efficiency function

h   centrifuge disc separation distance, m

I	ionic strength, $kmol\ m^{-3}$
I_s	intensity of segregation, Eq. (31)
k_D	constant in density Eq. (36)
k	Boltzmann constant, $1.3805 \times 10^{-23}\ JK^{-1}$
K_A	rate constant, Eq. (9), $m^3\ s^{-1}$
K	constant in solubility equations
N	particle number concentration, m^{-3}
N_s	stirrer speed, rps.
Q	centrifuge feed rate, $m^3\ s^{-1}$
q	net surface charge on a molecule, C
r	radial position in centrifugal field, m
R	gas constant, $8.314\ Jmol^{-1}\ K^{-1}$
Re_p	particle Reynolds number $= \varrho_f dv/\mu$
S	solubility, $kg\ m^{-3}$
t	time, s
T	absolute temperature, K
v	velocity, $m\ s^{-1}$

Greek symbols

α	collision effectiveness factor, Eq. (17)
β, β'	constants in solubility equations
$\Delta\varrho$	density difference: $\varrho_a - \varrho_f$, $kg\ m^{-3}$
ε	energy dissipated per unit mass, $W\ kg^{-1}$
η	turbulent microscale, m
θ	included angle between centrifuge axis and disc surface
μ_i	chemical potential of species i, $Jmol^{-1}$
μ	dynamic viscosity, $Ns\ m^{-2}$
v	kinematic viscosity, $m^2\ s^{-1}$
ϱ_a	aggregate density, $kg\ m^{-3}$
ϱ_f	fluid density, $kg\ m^{-3}$
$\varphi(h)$	potential energy of interaction between two particles at distance h apart, J
φ_v	volume fraction of particles in suspension
ω	angular velocity, $rad\ s^{-1}$

10 References

1. Rothstein, F., Rosenoer, V. M., Hughes, W. L.: Albumin purification, In: Albumin; Structure, Function and Uses in Man (Rosenoer, V. M., Rothschild, M. A. (eds.)), p. 7. New York: Pergamon Press 1977
2. Kuntz, O. D., Kauzmann, W.: Adv. Protein Chem. 28, 239 (1974)
3. Ives, K. J. (ed.): The Scientific Basis of Flocculation, Alphen aan den Rijn, Sijthoff and Noordhoff 1978
4. Morris, C. J. O. R., Morris, P.: Separation Methods in Biochemistry, p. 834. London: Pitman 1976[2]
5. Edsall, J. T.: Adv. Protein Chem. 3, 383 (1947)
6. Lyklema, J.: The Scientific Basis of Flocculation (Ives, K. J. (ed.)), p. 3. Alphen aan den Rijn, Sijthoff and Noordhoff 1978

7. Shaw, D. J.: Introduction to Colloid and Surface Chemistry, p. 137. London: Butterworths 1966
8. Prieve, D. C., Ruckenstein, E.: J. Colloid and Interface Sci. 73, 539 (1980)
9. Nozaki, Y., Tanford, C.: Methods in Enzymology (Hirs, D. H. W. (ed.)), 11, 715 (1967)
10. McBain, J. W.: Colloid Science, Lexington, Mass., D.C. Heath and Co. 1950
11. Whitney, R. M.: Food Colloids (Graham, H. E. (ed.)), p. 1. Westport Conn., Avi 1977
12. Riddick, T. M.: Control of Colloid Stability through Zeta Potential, Pennsylvania: Livingstone Publ. Co. 1968
13. Payens, J.: J. Dairy Res. 46, 291 (1979)
14. Kirchmeier, O.: Netherlands Milk and Dairy J. 27, 191 (1973)
15. Darling, D. F., Dickson, J.: J. Dairy Res. 46, 329 (1979)
16. Cohn, E. J.: Physiol. Rev. 5, 349 (1935)
17. Jirgensons, B.: Organic Colloids, p. 231. Amsterdam: Elsevier 1958
18. von Hippel, P. H., Peticolas, V., Schack, L., Karlson, L.: Biochem. 12, 1256 (1973)
19. Melander, W., Horvath, C.: Arch. Biochem. Biophys. 183, 200 (1977)
20. Salt, D., Dunnill, P., Leslie, R. B., Lillford, P. J.: Eur. J. Appl. Microbiol. Biotechnol. 14, 144 (1982)
21. Frigerio, N. A., Hettinger, T. P.: Biochim. Biophys. Acta 59, 228 (1962)
22. Schubert, P. F., Finn, R. K.: Biotech. Bioeng. 23, 2569 (1981)
23. Pennell, R. B.: Fractionation and isolation of purified components by precipitation methods, In: The Plasma Proteins (Putnam, F. W. (ed.)), Vol. 1, p. 9. New York: Academic Press 1960
24. Iverius, P. H., Laurent, T. C.: Biochim. Biophys. Acta 133, 371 (1967)
25. Ogston, A. G.: Biochem. J. 117, 85 (1970)
26. Albertsson, P. A.: Adv. Protein Chem. 24, 309 (1970)
27. Kroner, K. H., Kula, M. R.: Process Biochem. 13 (4), 7 (1978)
28. Foster, P. R., Dunnill, P., Lilly, M. D.: Biochim. Biophys. Acta 317, 505 (1973)
29. Miekka, S. I., Ingham, K. C.: Arch. Biochem. Biophys. 191, 525 (1978)
30. Ingham, K. C.: ibid. 186, 106 (1978)
31. Busby, T. F., Ingham, K. C.: Vox Sang. 39, 93 (1980)
32. Ledward, D. A.: Protein-polysaccharide interactions, In: Polysaccharides in food (Blanshard, J. M. V., Mitchell, J. R. (eds.)), London: Butterworths 1978
33. Imeson, A. P., Ledward, D. A., Mitchell, J. R.: J. Sci. Fd. Agric. 28, 661 (1977)
34. Gekko, K., Noguchi, H.: J. Agric. Food Chem. 26, 1409 (1978)
35. Oncley, J. L., Walton, K. W., Cornwall, D. G.: J. Amer. Chem. Soc. 79, 4666 (1957)
36. Hill, R. D., Zadow, J. G.: New Zealand J. Dairy Sci. Technol. 13, 61 (1978)
37. Vetter, H., Thum, W., Naeher, G.: 9th FEBS Meet., Dublin, April 1973
38. UK Pat. GB 1298431, 1972
39. Sternberg, M., Herschberger, D.: Biochim. Biophys. Acta 342, 195 (1974)
40. Greenwalt, T. J., Steane, E. A., Van Oss, C. J.: Fed. Proc. 25, 612 (1966)
41. Ishii, S., Hirata, A., Watanabe, T.: J. Jap. Soc. Food Sci. Technol. 26, 279 (1979)
42. Sternberg, M. Z.: Biotech. Bioeng. 12, 1 (1970)
43. Cerbulis, J.: J. Agric. Food Chem. 26, 806 (1978)
44. Gurd, F. R. N., Wilcox, P. E.: Adv. Protein Chem. 11, 311 (1956)
45. Dixon, M., Webb, E. C.: ibid. 16, 197 (1961)
46. Czok, R., Bucher, T.: ibid. 15, 315 (1960)
47. Curling, J. M. (ed.): Methods of Plasma Protein Fractionation, p. 3, New York: Academic Press 1980
48. Honig, W., Kula, M. R.: Anal. Biochem. 72, 502 (1975)
49. Alameri, E.: Suomen Kemistiehti 28B, 28 (1955)
50. Smoluchowski, M.: Z. Physik. Chem. 92, 155 (1917)
51. Ives, K. J.: The Scientific Basis of Flocculation (Ives, K. J. (ed.)), p. 37. Alphen aan den Rijn, Sijthoff and Noordhoff 1978
52. Higuchi, W. I. et al.: J. Pharm. Sci. 52, 49 (1963)
53. Turkevich, J.: Am. Scientist 47, 97 (1959)
54. Fuchs, N. A.: The Mechanics of Aerosols, p. 305. Pergamon Press 1964
55. Spielman, L. A.: The Scientific Basis of Flocculation (Ives, K. J. (ed.)), p. 63. Alphen aan den Rijn, Sijthoff and Noordhoff 1978

56. Okuyama, K., Kousaka, Y., Payatakes, A. C.: J. Colloid and Interface Sci. *81* (1), 21 (1981)
57. Dalgleish, D. G., Brinkhuis, J., Payens, T. A. J.: Eur. J. Biochem. *119*, 257 (1981)
58. Dalgleish, D. G.: Biophysical Chem. *11*, 147 (1980)
59. Parker, T. G., Dalgleish, D. G.: Biopolymers *16*, 2533 (1977)
60. Dalgleish, D. G., Parker, T. G.: J. Dairy Research *46*, 259 (1979)
61. Holt, C.: Biochim. Biophys. Acta *400*, 293 (1975)
62. Tanford, C.: Physical Chem. of Macromolecules, New York: Wiley 1961
63. Payens, T. A. J., Wiersman, A. K.: Biophysical Chem. *11*, 137 (1980)
64. Slattery, C. W.: ibid. *6*, 59 (1977)
65. Horne, D. S.: J. Dairy Research *46*, 265 (1979)
66. Ho, N. F. H., Higuchi, W. I.: J. Pharm. Sci. *56*, 248 (1967)
67. Hoare, M.: Trans. I. Chem. E. *60* (2), 79 (1982)
68. Hoare, M.: ibid. *60* (3), 157 (1982)
69. Swift, D. L., Friedlander, S. K.: J. Colloid Sci. *19*, 621 (1964)
70. Mason, S. G.: J. Colloid and Interface Sci. *58*, 275 (1977)
71. van de Ven, T. G., Mason, S. G.: Colloid and Polymer Sci. *255*, 468 (1977)
72. Gregory, J.: Chem. Eng. Sci. *36*, 1789 (1981)
73. Saffman, P. G., Turner, J. S.: J. Fluid Mech. *1*, 16 (1956)
74. Camp, T. R., Stein, P. C.: J. Boston Soc. Civ. Engrs. *30*, 219 (1943)
75. Tomi, D. T., Bagster, D. F.: Trans. I. Chem. E. *56*, 1 (1978)
76. Argaman, Y., Kaufman, W. J.: Proc. A.S.C.E. Sanit. Eng. Div. *96*, 223 (1970)
77. Gregory, J.: The Scientific Basis of Flocculation (Ives, K. J. (ed.)); p. 101. Alphen aan den Rijn, Sitjhoff and Noordhoff 1978
78. Walles, W. E.: J. Colloid and Interface Sci. *27*, 797 (1968)
79. Napper, D. H.: ibid. *32*, 106 (1970)
80. Glasgow, L. A., Luecke, R. H.: Ind. Eng. Chem. Fundam. *19*, 148 (1980)
81. Parker, D. S., Kaufman, W. J., Jenkins, D.: Proc. A.S.C.E., Sanit. Eng. Div. *98*, SA1, 79 (1972)
82. Choulalogou, C. A., Tavlarides, L. L.: Chem. Eng. Sci. *32*, 1289 (1977)
83. Hinze, J. O.: A. I. Ch. E. J. *1*, 289 (1955)
84. Shinnar, R.: J. Fluid Mech. *10*, 259 (1961)
85. Konno, M. et al.: J. Chem. Eng. Japan *13*, 67 (1980)
86. Nienow, A. W., Conti, R.: Chem. Eng. Sci. *33*, 1077 (1978)
87. Conti, R., Nienow, A. W.: ibid. *35*, 543 (1980)
88. Karabelas, A. J.: A. I. Ch. E. J. *22*, 765 (1976)
89. Hoare, M. et al.: Ind. Eng. Chem. accepted for publication, 1982
90. Smith, D. K. W., Kitchener, J. A.: Chem. Eng. Sci. *33*, 1631 (1978)
91. Twineham, M. et al.: (Dept. Chem. and Biochem. Eng., Univ. College London): to be published, 1982
92. Bell, D. J., Dunnill, P.: Biotech. Bioeng. *24*, 1271 (1982)
93. Tambo, N., Hozumi, H.: Water Res. *13*, 421 (1979)
94. Thomas, D. G.: A. I. Ch. E. J. *10*, 517 (1964)
95. Akers, R. J., Machin, G. P.: The Second World Filtration Congress, London, p. 365, (18—20 Sept. 1979)
96. Huck, P. M., Murphy, K. L.: Proc. A.S.C.E., Env. Eng. Div. *104*, 767 (1978)
97. Matsuo, T., Unno, H.: ibid. *107*, 527 (1981)
98. Tomi, D. T., Bagster, D. F.: Trans. I. Chem. E. *56*, 9 (1978)
99. Virkar, P. D. et al.: Biotech. Bioeng. *24*, 871 (1982)
100. Maruscak, A., Baker, C. G. J., Bergougnou, M. A.: The Canadian J. Chem. Eng. *49*, 819 (1971)
101. Baker, C. G. J., Bergougnou, M. A.: ibid. *52*, 246 (1974)
102. Wahl, E. F., Baker, C. G. J.: ibid. *49*, 742 (1971)
103. Ives, K. J., Bhole, A. G.: Proc. A.S.C.E. Env. Eng. Div. *99*, EE1, 17 (1973)
104. Tambo, N., Watanabe, Y.: Water Res. *13*, 409 (1979)
105. Hunter, R. J., Frayne, J.: J. Colloid and Interface Sci. *76*, 107 (1980)
106. Mullin, J. W.: Crystallization, p. 339. Butterworths, 1971[2]
107. Nagata, S.: Mixing, New York, J. Wiley., 1975

108. Uhl, V. W., Gray, J. B.: Mixing: Theory and Practice, Vol. 1, New York, Academic Press, 1966
109. Levenspiel, O.: Chemical Reaction Eng., New York, J. Wiley and Sons Inc., 1972
110. Overbeek, J. Th. G.: J. Colloid and Interface Sci. *58*, 408 (1977)
111. Southward, C. R., Aird, R. M.: New Zealand J. Dairy Sci. *13*, 77 (1978)
112. Foster, P. R., Dunnill, P., Lilly, M. D.: Biotech. Bioeng. *18*, 545 (1976)
113. Nauman, E. B.: J. Macromol. Sci. Revs. Macromol. Chem. *C10* (1), 75 (1974)
114. Bell, D. J. et al.: Biotech. Bioeng. *24*, 127 (1982)
115. Purcell, E. M.: J. Fluid Mech. *84* (3), 551 (1978)
116. Brodkey, R. S.: Chem. Eng. Commun. *8* (1–3), 1 (1981)
117. Patterson, G. K.: ibid, *8* (1–3), 25 (1981)
118. Van t'Riet, K., Smith, J. M.: Chem. Eng. Sci. 30, 1093 (1975)
119. Vrale, L., Jordan, R. M.: J.A.W.W.A. *63* (1), 52 (1971)
120. Bratby, J.: Coagulation and Flocculation, England, Uplands Press Ltd. 1980
121. Camp, T. R.: Trans. A.S.C.E. *120*, 1 (1955)
122. Leentvaar, J., Ywema, T. S. J.: Water Res. *14*, 135 (1980)
123. Kitchener, J. A., Gochin, R. J.: ibid. *15*, 585 (1981)
124. Bell, D. J., Dunnill, P.: Biotech. Bioeng. accepted for publication 1981
125. O'Melia, C. R.: The Scientific Basis of Flocculation (Ives, K. J. (ed.)), p. 219. Alphen aan den Rijn, Sijthoff and Noordhoff 1978
126. Pielkenrood-Vinitex, B. V.: UK Pat. GB 1433171, 1976
127. Hiedemann, E., Brandt, O.: UK Pat. GB 508675, 1937
128. Obiakor, E. K., Whitmore, R. L.: Nature (London) *205*, 381 (1965)
129. Anada, H. R., Shah, Y. T., Klinzing, G. E.: The Canadian J. Chem. Eng. *52*, 715 (1974)
130. Jewett, W. R.: US Pat. 3826740, 1974
131. Foster, P. R.: Proc. Int. Workshop Techn. Protein Separation Impr. Blood Plasma Fract. (Sandberg, H. (ed.)), USDHEW. Publ. No. (NIH) 78–1422, p. 54, Washington, D.C. 1978
132. Friedli, H. et al.: Vox Sang. *31*, 289 (1976)
133. Wolter, D.: Proc. Int. Workshop Techn. Protein Sepn. Impr. Blood Plasma Fract. (Sandberg, H. (ed.)), USDHEW Publ. No. (NIH) 78–1422, p. 129, Washington, D.C. 1978
134. Duncan, I. M., Holliday, A. G.: 2nd World Filtn. Congress, 215 (1979)
135. Nyiri, L.: Process Biochem. *4* (8), 27 (1969)
136. Underkofler, L. A.: Enzymes in Food Processing (Reed, G. (ed)), p. 217. New York: Academic Press 1975[2]
137. Schweitzer, P. A.: Handb. of Separation Techniques for Chemical Engineers, p. 4, New York: McGraw Hill 1979
138. Somasundaran, P.: Sepn. and Purifn. Methods *1* (1), 117 (1972)
139. Iibuchi, S., Chiang, W., Yano, T.: J. Agric. Biol. Chem. *44* (8), 1811 (1980)
140. Southward, C. R., Walker, N. J.: New Zealand J. Dairy Sci. Tech. *15*, 201 (1980)
141. Beaton, N. C.: Ultrafiltration Membranes and Applications (Cooper, A. R. (ed.)), New York: Plenum Press 1980
142. Knuckles, B. E. et al.: J. Agric. Food Chem. *28*, 32 (1980)
143. Bungay, H. R., Millspaugh, M. P.: 2nd Chem. Congr. N. Amer. Count: Div. of Microbiol. and Biochem. Tech. 1980
144. Wang, D. I. C. et al.: Fermentation and Enzyme Techn., p. 295. New York: J. Wiley and Sons 1979
145. I. Chem E.: User Guide for The Safe Operation of Centrifuges, Rugby, England, I. Chem. E. 1979
146. Charm, S., Wong, B. L.: Enzyme Microb. Techn. *3* (2), 111 (1981)
147. Foster, P. R., Watt, J. G., Dickson, A. J.: A. I. Ch. E. Symp., Processing and Fractionation of Blood Plasma, Philadelphia, June 8–12 (1980)
148. Hemfort, H.: Proc. Int. Workshop Techn. Protein Sepn. Impr. Blood Plasma Fract. (Sandberg, H. (ed.)), USDHEW Publ. No (NIH) 78–1422, p. 81, Washington, D.C. 1978
149. Allen, T.: Particle Size Measurement, p. 301. London: Chapman and Hall 1981
150. Kuntz, I. D., Crippen, G. M.: Int. J. Peptide Protein Res. *13*, 223 (1979)
151. Ifft, J. B.: Biophys. Chem. *5*, 137 (1976)
152. Lagvankar, A. L., Gemmell, R. S.: J.A.W.W.A. *9*, 1040 (1968)

153. Munro, P. A.: New Zealand Dairy Sci. Techn. *15*, 225 (1980)
154. Higgins, J. J. et al.: Biotech. Bioeng. *20*, 159 (1978)
155. Trim, A. R.: Biochem. J. *73*, 298 (1959)
156. Atkinson, A., Jack, G. W.: Biochim. Biophys. Acta *308*, 41 (1973)
157. Oxenburgh, M. S., Shoswell, A. M.: Nature (London) *204*, 1416 (1965)
158. Melling, J., Atkinson, A.: J. Appl. Chem. Biotechnol. *22*, 739 (1972)
159. Russell, A. J.: PhD Thesis, Univ. of London 1981
160. Vinogradova, M. G. et al.: J. Appl. Chem. (USSR) *44* (5), 1062 (1971)
161. Brenner, H., in: Progress in Heat and Mass Transfer (Schowalter, W. R. et al., (eds.)), Vol. 5, p. 89. Pergamon Press 1970
162. Mewis, J., Spaull, A. J. B.: Adv. Colloid and Interface Sci. *6*, 173 (1976)
163. Michaels, A. S., Bolger, J. C.: Ind. Eng. Chem. Fundam. *1* (3), 153 (1962)
164. Thomas, D. G.: A. I. Ch. E. J. *9* (3), 310 (1963)
165. Firth, B. A., Hunter, R. F.: J. Colloid and Interface Sci. *57*, 248 (1976)
166. Van den Tempel, M.: ibid. *71* (1), 18 (1979)
167. Strivens, T. A.: ibid. *57* (3), 476 (1976)
168. Lee, C. H., Rha, C.: Food Texture and Rheology (Sherman, P., (ed.)), p. 245. New York: Academic Press 1979
169. Meier, P. M.: Proc. Int. Workshop Techn. Protein Sepn. Impr. Blood Plasma Fract. (Sandberg, H. (ed.)), USDHEW Publ. No (NIH) 78–1422, p. 118, Washington, D.C. 1978
170. Lee, C. H., Rha, C.: J. Food Sci. *43*, 79 (1978)
171. Vu, J. T., Munro, P. A.: 9th Australasian Conf. Chem. Eng., Christchurch N.Z., Aug. 30, 1981
172. Belter, P. A., Beckel, A. C., Smith, A. K.: Ind. Eng. Chem. *36* (9), 799 (1944)
173. Edwards, R. H. et al.: J. Agric. Food Chem. *23* (4), 620 (1975)
174. Ambler, C. M.: Chem. Eng. Progr. *48* (3), 150 (1952)
175. Jury, S. H., Locke, W. L.: A. I. Ch. E. J., *3* (4), 480 (1957)
176. Frampton, G. A.: Chem. Proc. Eng., p. 402, Aug. 1963
177. Sokolov, V. I.: Moderne Industriezentrifugen, Berlin, Verlag Technik 1967
178. Baron, G., Wajc, S.: Chem.-Ing.-Tech. *51* (4), 333 (1979)
179. Trowbridge, M. E. O'K.: The Chem. Eng. *162*, A73 (1962)
180. Purchas, D. B.: Solid-liquid Separation Equipment Scale-up, England, Uplands Press Ltd. 1977
181. Svarovsky, L.: Solid-Liquid Separation. (Svarovsky, L. (ed.)), p. 31. London: Butterworths 1977
182. Gibson, K. R.: Proc. Symp. Solid-Liquid Separation Pract. *153*, 1 (1979)
183. Sokolov, V. I., Borisenko, A. A.: Soviet Chem. Ind. *9* (6), 486 (1977)
184. Gupta, S. K.: Chem. Eng. J. *22*, 43 (1981)
185. Zeitsch, K.: Trans. I. Chem. E. *56*, 281 (1978)
186. Sokolov, V. I., Gorbunova, V. V., Rusakova, A. A.: Chem. Petrol. Eng. *11* (1), 28 (1975)
187. Golovko, Yu. D.: Theor. Found. Chem. Eng. *7* (4), 583 (1973)
188. Gol'din, E. M.: ibid. *5* (2), 246 (1971)
189. Zastrow, J.: Int. Chem. Eng. *16* (3), 515 (1976)
190. Carlsson, C.-G.: Ph. D. Dissertation, Chalmers Univ. of Techn., Gothenburg 1979
191. Brunner, K.-H., Molerus, O.: Ger. Chem. Eng. *2*, 228 (1979)
192. Vinogradova, M. G., Romankov, P. G.: Theor. Found. Chem. Eng. *11* (6), 780 (1977)
193. Sokolov, V. I., Semenov, V. V., Gorbinova, V. V.: ibid. *11* (2), 220 (1977)
194. Skvortsov, L. S.: ibid. *10* (6), 833 (1976)
195. Vinogradova, M. G., Plyushkin, S. S., Romankov, P. G.: ibid. *12* (2), 202 (1978)
196. Chesnokov, V. M., Tutevich, V. P., Mizeretskii, N. N.: ibid. *9* (1), 79 (1975)
197. Skvortsov, L. S., Rachitskii, V. A.: ibid. *12* (4), 526 (1978)
198. Sokolov, V. I., Semenov, E. V.: J. Appl. Chem. USSR *49* (7), 1575 (1976)
199. Semenov, E. V., Karamzin, V. A.: Theor. Found. Chem. Eng. *13* (5), 632 (1979)
200. Belovolov, N. V., Borts, M. A., Gupalo, Yu. P.: ibid. *11* (4), 493 (1977)
201. Zurcher, H. E.: Filtr. Sepn. *11* (6), 606 (1974)
202. Gösele, W.: Ger. Chem. Eng. *3*, 353 (1980)
203. Redeker, D.: Chem.-Ing.-Tech. *52* (6), 542 (1980)
204. Stahl, W.: ibid. *47* (20), 853 (1975)

205. Vallet, L.: J. Biochem. Microbiol. Tech. Eng. *2* (2), 121 (1960)
206. Miller, G., Mizrahi, A., Shawit, A.: Biotech. Bioeng. *10* (5), 684 (1968)
207. Hoare, M.: Ph. D. Thesis, Univ. of London 1981
208. United Kingdom Atomic Energy Authority: UK Pat. GB 2005163, 1979
209. Gullotta, J. D., Boadway, J. D.: Water Polln. Control *117* (6), 58 (1979)
210. Lagergren, B.: Int. Congr. Eng. and Food, 1979[2]
211. Westfalia Separator A.G.: UK Pat. GB 1500622, 1978
212. Westfalia Separator A.G.: W. German Pat. No. 2631110, 1978
213. Bott, E. W.: Filtr. Sepn. *17* (5), 426 (1980)
214. Lagergren, B.: Food Process Eng., Vol. 1 (Linko, P. et al. (eds.)), New York: Appl. Science Publ. Ltd. 1980
215. Alfa-Laval, AB: UK Pat. Application, GB 2052315A, 1981
216. Alfa-Laval, AB: W. German Pat. No. 2845733, 1979
217. Pennwalt Corp.: UK Pat. GB 1408997, 1975
218. Kloeckner-Humboldt-Dentz, AG: UK Pat. GB 2015386, 1979
219. Hsu, H. W.: Sepn. and Purifn. Methods. *5* (1), 51 (1976)
220. Charm, S. E., Matteo, C. C.: Method Enzymol. *22*, 476 (1971)
221. Cline, G. B.: Progr. Sepn. Purifn. p. 297 (Perry, E. S., Van Oss, C. J. (eds.)), Vol. 4, New York: Wiley Intersc. 1971
222. De, S. S.: Techn. of Production of Edible Flours and Protein Products from Soybean, Agricult. Services Bull. No. 11, p. 112, Food and Agric. Org. United Nations, Rome 1974[2]
223. Foster, P. R., Watt, J. G.: The CSVM Fractionation Process, In: Methods of Plasma Protein Fractionation. (Curling, J. M. (ed.)), p. 17, New York: Academic Press 1980
224. Dunnill, P., Lilly, M. D.: Biotech. Bioeng. Symp. *3*, 97 (1972)
225. Dunnill, P., Currie, J. A., Lilly, M. D.: Biotech. Bioeng. *12*, 63 (1970)
226. Aunstrup, K.: Industrial Approach to Enzyme Production, In: Biotechn. Applications of Proteins and Enzymes (Bohak, Z., Sharon, N. (eds.)), p. 44, New York: Academic Press 1977
227. Murkes, J., Carlsson, C. G.: Filtration and Separation, p. 18. Jan./Feb. 1978
228. Virkar, P. D. et al.: Biotech. Bioeng. *23*, 425 (1981)

Ultrafiltration for the Separation of Biocatalysts

Erwin Flaschel
Institut de Génie Chimique, Ecole Polytechnique Fédérale de Lausanne, Ecublens
CH-1015 Lausanne, Switzerland

Christian Wandrey
Institut für Biotechnologie, Kernforschungsanlage Jülich, D-5170 Jülich, FRG

Maria-Regina Kula
Gesellschaft für Biotechnologische Forschung, D-3300 Braunschweig-Stöckheim, FRG

The application of ultrafiltration in biotechnology is reviewed emphasizing the separation of catalytically active species. Ultrafiltration as a separation process as well as its application to membrane reactors is analyzed. On the basis of an application-oriented theory of ultrafiltration, the essential aspects for process design are described. A survey of applications is given.

Ultrafiltration has proved to be a very versatile separation process for biocatalysts owing to the possibility of independently adjusting the temperature, the avoidance of phase transition and the low energy requirement. It will be shown that ultrafiltration devices, suitable for the isolation of biocatalysts, can also be applied efficiently for their re-use in catalytic processes. The main advantages of employing biocatalysts in ultrafiltration membrane reactors are that continuous operation is possible in homogeneous phase and that immobilization know-how is not required.

Future trends can be predicted with respect to the development of sterilisable membranes with improved narrow pore size distribution and with surfaces that will not affect fragile biocatalysts. Continuous coenzyme regeneration in membrane reactors and biomass recycling in continuous fermentation processes, in order to uncouple the retention time of the catalyst from the hydraulic retention time will result in increased application.

1 Introduction

Since biotechnology has been attributed a key-role in solving many of our future problems there is a demand for both new and more efficient biological processes, and the technology which leads to the economic use of its potential. Technology includes all parts of a process and it is widely known that the main operating costs are frequently found in the downstream processes for product separation.

The recovery of biological products requires separation processes, which meet the specific requirements demanded by biological systems. For the separation of bio-catalysts, enzymes and microorganisms, their high molar mass and temperature sensitivity have to be taken into account. In addition, the separation might also require aseptic conditions. Such requirements can be satisfied by ultrafiltration, which is especially suited to the recovery of substances of relatively high molar mass.

1.1 Ultrafiltration as a Separation Process

Ultrafiltration is a separation process, in which the selectivity is mainly determined by the particle size. It differs from conventional filtration and microfiltration in that the separation takes place at the molecular level, which implies that the particles to be retained are mostly soluble. Ultrafiltration is characterized by the separation of particles with a molar mass within the range of 500 to 500,000 g mol^{-1}. This requires pore diameters of between 1 and 20 nm. There is no distinct demarcation between ultrafiltration and reverse osmosis (hyperfiltration). Both reverse osmosis and ultra-filtration are pressure driven processes with enforced flow through essentially porous synthetic membranes. However, the selectivity of reverse osmosis membranes depends on the interaction of the molecules with the polymer matrix and cannot be solely explained by the membrane's porous structure. In reverse osmosis, a minimum pressure has to be applied to obtain a permeate (normally solvent), depending on the osmotic backpressure, which in turn depends on the concentration of retained low molar mass solutes. Since ultrafiltration is mainly used for the separation of high molar mass substances, the osmotic backpressure can normally be neglected. Ultra-filtration must also be distinguished from dialysis, a process characterized by diffusive mass transport.

The separation of macrosolutes by ultrafiltration has found many applications and increasing interest in quite different areas [1], since it exhibits several advantages over common separation processes. These are:

— low energy requirement
— independently adjustable operating temperature
— no phase transition
— sterile conditions can be achieved
— it lowers the consumption of chemical agents e.g. for precipitation
— the possibility of batch and continuous operation
— relatively uncritical scale-up.

Therefore, ultrafiltration is most commonly used as a relatively inexpensive separation step before applying other costly methods of further purification or con-centration.

1.2 Application of Ultrafiltration in Biotechnology

The first membranes with reasonable hydraulic permeabilities for pressure driven separation processes were reported by Sourirajan and Loeb [2] in the early 1960's. The development started with desalination membranes and was extended to those retaining macrosolutes, especially for application in sewage treatment [3]. The break-through came from the development of asymmetric membranes, which are composed of an ultra-thin microporous layer, supported by a macroporous structure. In comparison to symmetrically structured membranes, asymmetric ones exhibit a much higher permeability owing to a reduced effective pore length and the avoidance of deep bed filtration. The asymmetric ultrafiltration membrane is now standard and is manufactured in numerous geometric configurations and modules, so as to achieve high surface areas per unit volume. The modules range from flat membranes to plate-and-frame systems [4], sandwich modules, tubular systems, capillary devices and hollow fiber cartridges [5, 6, 7].

By 1965, the first laboratory scale ultrafiltration membranes were commercially available [3]. At that time, protein concentration by means of ultrafiltration membranes was reported by Blatt et al. [8]. In 1968, Michaels [9] suggested continuous ultrafiltration for the retention of biocatalysts in a continuously operated membrane reactor, and the first application of such a system was reported in the same year by Blatt et al. [10].

The main application of ultrafiltration in biotechnology is for the concentration of extracellular enzymes from dilute solutions. Of similar importance is the removal of low molar mass compounds (e.g. salts) from enzyme solutions. This is achieved by feeding pure water or buffer in enforced flow through an ultrafiltration device, so as to replace the solution of low molar mass molecules and thereby retain the enzyme. This process is called diafiltration in analogy to dialysis. As far as the fractionation of high molar mass products is concerned, there have been attempts to remove poly-meric substances of either higher or lower molar mass than that of the desired enzyme by means of ultrafiltration. Such a fractionation of biopolymers seems to be a desirable secondary effect during concentration and diafiltration.

The harvesting of biomass, especially of bacteria und cellular debris by so called "crossflow filtration", seems to find increasing industrial interest.

Since biocatalysts can be effectively separated from low molar mass impurities by ultrafiltration, it is also possible to separate the catalysts from low molar mass reaction products. Thus homogeneous catalysis in continuous membrane reactors becomes possible. This principle has been widely used, especially with hydrolases, and has been applied on an industrial scale [11]. The membrane reactor concept is of particular interest for multi-enzyme systems with continuous cofactor regeneration. For this purpose the cofactor has to be enlarged in order to be retained together with the enzymes involved. This has been achieved by covalent coupling of the cofactor to water soluble polymers.

Resting cells can also be used as biocatalysts in membrane reactors. As far as continuous fermentation is concerned, washout of cells can be avoided by downstream separation and recycling of biomass. Thus higher productivities are achieved in response to the higher biocatalyst concentrations present.

2 Theory of Ultrafiltration

The main parameters influencing ultrafiltration are the properties of the membrane and of the medium as well as the hydrodynamic conditions at the membrane. The economically important criteria are solvent permeability, solute retention and membrane stability.

Ultrafiltration theory will be described here by means of the pore model. On this basis guidelines are derived for the application of ultrafiltration. The separation of biocatalysts and their use in membrane reactors are discussed separately. Ultrafiltration devices for concentration purposes are normally operated at constant pressure and at rather high catalyst concentrations, while in continuously operated membrane reactors a constant transmembrane flow rate is applied using comparatively small catalyst concentrations. While in the first case the chief aim is directed to high solute concentrations, in the second case, extremely high catalyst retention is the objective.

2.1 Main Parameters

Since ultrafiltration is based on forced convection through porous membranes, a physical description of solvent permeability and solute rejection is needed. The performance of ultrafiltration membranes is normally described by means of the pore model. It assumes ideal cylindric pores normal to the membrane surface. The solvent can pass through these pores while solutes are more or less retained. The flux (transmembrane flow rate per unit area) across a membrane depends on a transmembrane pressure gradient as the driving force. For pure solvent or negligible solute concentrations this interrelation can be expressed by Hagen-Poiseuille's law:

$$J = \varepsilon_m \frac{d_p^2}{32 \eta l_p} \Delta p \tag{1}$$

where

J	$m^3\ m^{-2}\ s^{-1}$	flux (area-specific transmembrane flow rate)
ε_m	—	membrane porosity
d_p	m	pore diameter
η	Pa s	dynamic viscosity
l_p	m	pore length
Δp	Pa	transmembrane pressure drop

To characterize the permeability of a distinct membrane, the hydraulic resistance is useful and may be defined by:

$$J = \frac{1}{W_m} \Delta p \tag{2}$$

where

W_m	$Pa\ s\ m^{-1}$	hydraulic membrane resistance

This quantity is a characteristic value for a given membrane and is defined for pure solvent flux. It takes into account all deviations from the ideal assumptions of Hagen-Poisseuille's law. If there are solutes which are retained by the membrane, the hydraulic resistance increases significantly, even at low solute concentrations, due to adsorption to the membrane. At higher solute concentrations the hydraulic resistance rises further, due to a phenomenon called concentration polarization. The flux of the solvent drives the solute against the membrane and since it is retained, a higher solute concentration results in front of it. At steady state, a stable concentration gradient exists owing to backdiffusion of the concentrated solute. If the concentration and/or the flux is high enough, the saturation concentration of the solute will be attained at the membrane. This can lead to a build-up of a secondary membrane of solute particles with high hydraulic resistance:

$$J = \frac{1}{W_m + W_g} \Delta p \tag{3}$$

where

W_g Pa s m^{-1} hydraulic resistance of the secondary membrane (gel layer)

The hydraulic resistance of the membrane is supposed to be independent of flux, while that of the secondary membrane increases when the flux is raised. This is due to additional precipitation of solute molecules in front of the membrane. The effect of concentration polarization is shown schematically in Fig. 1. At steady state, the sum of molar fluxes of solute must be zero:

$$J_{ic} - J_{if} - J_{id} = 0 \tag{4}$$

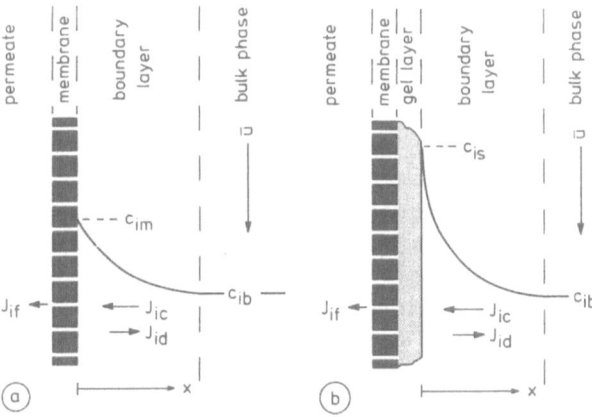

Fig. 1a and b. Schematic representation of concentration polarization without gel formation (**a**) and with gel formation (**b**)

where

J_{ic}	mol m^{-2} s^{-1}	molar flux of solute (convection)
J_{if}	mol m^{-2} s^{-1}	molar flux of solute (filtrate)
J_{id}	mol m^{-2} s^{-1}	molar flux of solute (back-diffusion)

Equation (4) can also be expressed in terms of solute concentration:

$$Jc_i - Jc_{if} + D_i \frac{dc_i}{dx} = 0 \tag{5}$$

where

c_i	mol m^{-3}	concentration of solute i
c_{if}	mol m^{-3}	solute concentration in the filtrate
D_i	m^2 s^{-1}	diffusion coefficient of solute i
$\dfrac{dc_i}{dx}$	mol m^{-4}	concentration gradient of solute i

To calculate the concentration profile of solute in the boundary layer, Eq. (5) can be integrated:

$$-\int_{c_{ib}}^{c_{im}} \frac{dc_i}{c_i - c_{if}} = \frac{J}{D_i} \int_{\delta}^{0} dx \tag{6}$$

where

c_{ib}	mol m^{-3}	solute concentration in the bulk phase
c_{im}	mol m^{-3}	solute concentration at the membrane
δ	m	boundary layer thickness
x	m	position before the membrane

leading to:

$$c_{im} = c_{if} + (c_{ib} - c_{if})\, e^{\left(J\frac{\delta}{D_i}\right)} \tag{7}$$

The boundary layer thickness (δ) depends on the hydrodynamic conditions. Normally δ is not exactly known, but as will be shown later, the ratio D_i/δ can be regarded as a mass-transfer coefficient (k_d).

Equation (7) can be solved for the flux:

$$J = k_d \ln \frac{c_{im} - c_{if}}{c_{ib} - c_{if}} \tag{8}$$

where

k_d	m s^{-1}	mass-transfer coefficient ($= D_i/\delta$)

For quantitative retention ($c_{if} = 0$) Eq. (8) is simplified to:

$$J = k_d \ln \frac{c_{im}}{c_{ib}} \qquad (9)$$

If the solute concentration at the membrane reaches the saturation concentration, c_{im} remains constant even if the bulk phase concentration is further increased:

$$J = k_d \ln \frac{c_{is}}{c_{ib}} \qquad (10)$$

where

c_{is} mol m^{-3} solute saturation concentration

From a plot of J versus ln (c_{ib}), the mass-transfer coefficient (k_d) can be obtained from the slope. It should be measured at constant transmembrane pressure and constant hydrodynamics (see Sect. 2.2).

To estimate the mass-transfer coefficient at different hydrodynamic conditions, the general boundary layer theory can be applied. Empirical correlations of dimensionless numbers (from heat and mass transfer theory) are employed. The most commonly used equations for laminar and turbulent flow are given for tubular membrane geometry in the form:

$$Sh = \frac{k_d d_t}{D_i} = f\,(Re, Sc, d_t/l_t) \qquad (11)$$

laminar flow:

$$Sh = 1.62 \left(Re\ Sc\ \frac{d_t}{l_t} \right)^{1/3} \qquad \text{Graetz [12], Lévêque [13]} \qquad (11\,a)$$

$$Sh = 0.664 \left(Re\ \frac{d_t}{l_t} \right)^{1/2} Sc^{1/3} \qquad \text{Gröber [14]} \qquad (11\,b)$$

turbulent flow:

$$Sh = 0.023\ Re^{0.8}\ Sc^{0.33} \qquad \text{Dittus, Boelter [15]} \qquad (11\,c)$$

$$Sh = 0.0096\ Re^{0.913}\ Sc^{0.345} \qquad \text{Harriot, Hamilton [16]} \qquad (11\,d)$$

where

Sh — Sherwood number
d_t m diameter of tubular membrane (otherwise equivalent hydraulic diameter)
Re — Reynolds number $\left(= \dfrac{\bar{u} d_t}{v} \right)$

$\bar{u}$	$m\ s^{-1}$	superficial fluid velocity in the flow channel along the membrane
v	$m^2\ s^{-1}$	kinematic viscosity
Sc	—	Schmidt number $\left(= \dfrac{v}{D_i}\right)$
l_t	m	length of tubular membrane

These equations can be solved to give the mass-transfer coefficient:

laminar flow:

$$k_d = 1.62 \left(\frac{\bar{u}D_i^2}{d_t l_t}\right)^{1/3} \tag{12a}$$

$$k_d = 0.664 \left(\frac{\bar{u}}{l_t}\right)^{1/2} \frac{D_i^{2/3}}{v^{1/6}} \tag{12b}$$

turbulent flow:

$$k_d = 0.023\, \frac{\bar{u}^{0.8} D_i^{0.67}}{d_t^{0.2} v^{0.47}} \tag{12c}$$

$$k_d = 0.0096\, \frac{\bar{u}^{0.913} D_i^{0.655}}{d_t^{0.079} v^{0.568}} \tag{12d}$$

For given systems, the mean linear velocity ($\bar{u}$) is the parameter which influences the mass-transfer coefficient. Under laminar flow conditions it should vary with the linear velocity to the power 0.3 to 0.5, whereas under turbulent conditions to the power 0.8 to 1. Turbulent conditions are normally assumed at Re > 2,000. For further design equations, especially those applied to stirred ultrafiltration devices, refer to Blatt et al. [17].

Deriving the gel-polarization model (Eq. (10)), it has been assumed that a gel layer is formed on the membrane. This model implies that the flux depends no longer on the membrane characteristics but almost exclusively on the hydraulic resistance of the gel layer. Therefore, the flux is closely related to the properties of the species forming the gel layer and the flow rate along the membrane. Another result of the model is that the flux becomes independent of pressure. Since the saturation concentration of solute is already attained at the membrane, the gradient for back-diffusion of solute particles cannot be further increased. Augmenting the transmembrane pressure will result in an initial increase of the flux. Owing to the higher flux, the gel layer thickness will increase, thereby increasing the hydraulic resistance until the former flux is re-established [17].

Therefore, it is evident that the flux can only be efficiently increased by means of enhancing the back-transport of the solute from the membrane into the bulk liquid phase. This includes all parameters which positively influence the mass-transfer coefficient (k_d).

The main parameters for ultrafiltration are listed in Table 1. Besides the diffusion coefficient of the solute (D_i) and the kinematic viscosity of the medium (v) which are constant for a given system, back-transport of retained species can be influenced mainly by the operating variables and the geometry of the ultrafiltration module. Among these, the superficial velocity of the medium has the greatest direct influence on mass transfer. Increasing the temperature facilitates mass transfer indirectly by lowering viscosity and enhancing the solute diffusion. Short flow channels are better than long ones (l_t) since the relative length where the laminar boundary layer is completely established can be decreased. The interrelation between length and diameter of the flow channel is complex and will be discussed later.

Table 1. Main parameters which influence the performance of ultrafiltration

	Membrane	Geometry	System	Operation
Basic para- meters	$\bar{d}_p$ σ_p^2 cut-off l_p ε_m	$d_t(d_h)$ l_t A_m	$d_i(M_i)$ σ_i^2 D_i v c_{is} c_{ib}	Δp J $\bar{u}$ T
Derived quantities	$W_m = f(\bar{d}_p, l_p, \varepsilon_m)$ $R_m = f(d_i)$ $R_i = 1 - c_{if}/c_{ib}$	$a = A_m/V$	$Sc = v/D_i$	$Re = \bar{u}d_t/v$ $Sh = f(Re, Sc, d_t/l_t)$ $k_d = Sh\, D_i/d_t$

Mass transfer can also be increased by means of so called turbulence promoters. The common idea is to disturb the velocity profile in the fluid passing along the membrane and thus decreasing the laminar sublayer. Motionless mixers [18, 19], wires adjacent to the membrane [20, 21] and fluidized beds of glass particles [22] have been employed as turbulence promoters.

The mathematical treatment of ultrafiltration flux as a function of all significant variables is possible for laminar flow and has to combine the velocity profile with the over-all balance equations, including diffusive back-transport of solute. It should further take into account that the gel layer thickness is a function of flow channel length and, therefore, the flux will vary along the membrane. The flux will also vary due to a pressure drop along the membrane. An entire treatment of this problem is given by Rauch [7], who also considers the effect of segregation of particles in laminar flow, i.e. the pinch effect. This latter effect involves a radial migration of solute particles owing to slip-spin magnus force and/or inertial effects (slip-shear force) due to flow in a shear field. It must be considered when particles such as colloids, microorganisms or latex are ultrafiltered (see Porter [6]).

The pinch effect can also be used to enhance mass transfer by introducing particles of suitable diameter into the ultrafiltration unit [23-25].

Another problem which leads to deviation of results from prediction by simple modelling is that some parameters such as the diffusion coefficient, the viscosity and the density depend on solute concentration [26]. If this is considered, better agreement between theory and experiment can be expected [21, 27-29].

If the solute molecules are not quantitatively retained by the membrane, their loss across the membrane should be proportional to their concentration at the membrane (c_{im}). Assuming constant performance of the membrane, the intrinsic retention is defined as:

$$R_m = 1 - \frac{c_{if}}{c_{im}}$$

(13)

where

R_m —   intrinsic retention coefficient

Theoretically, R_m is constant for a distinct membrane/solute combination and independent of flow conditions. Since c_{im} cannot be measured directly in practice, a flow dependent retention is obtained:

$$R = 1 - \frac{c_{if}}{c_{ib}}$$

(14)

where

R —   apparent retention coefficient

The retention of solute molecules by a membrane is determined by the mean pore size and the pore size distribution. The influence of other variables is rather complex. Since the intrinsic retention is assumed to be constant, the apparent retention depends on the solute concentration at the membrane (c_{im}). Therefore, theoretically, all the measures which tend to increase mass transfer should also lead to a better solute retention (for details, see Chapter 2.2).

2.2 Separation

The separation of solutes from solvent (concentration) and/or molecules of low molar mass (diafiltration) is normally achieved at constant pressure. The discussion can be simplified if at first quantitative retention of solute is assumed.

The flux across an ultrafiltration membrane is determined by the average trans-membrane pressure. For a pure solvent, the interrelation of both quantities is normally found to be linear. When the pressure is increased at a given solute concentration, this relationship is no longer linear. As schematically shown in Fig. 2, the deviation intensifies with increasing solute concentration. At high concentrations a kind of saturation curve results. This is due to the fact that at high pressure drop and solute concentration, the saturation of solute is attained at the membrane. The flux becomes independent of pressure as described in 2.1. This implies that it is favourable to work near the maximal allowable pressure drop since the highest flux can be achieved independently of solute concentration. Working at constant flux would mean that during a concentration step, the pressure would rise. Therefore, it would only be possible to work at the lowest flux which is determined by the pressure drop at the desired final solute concentration.

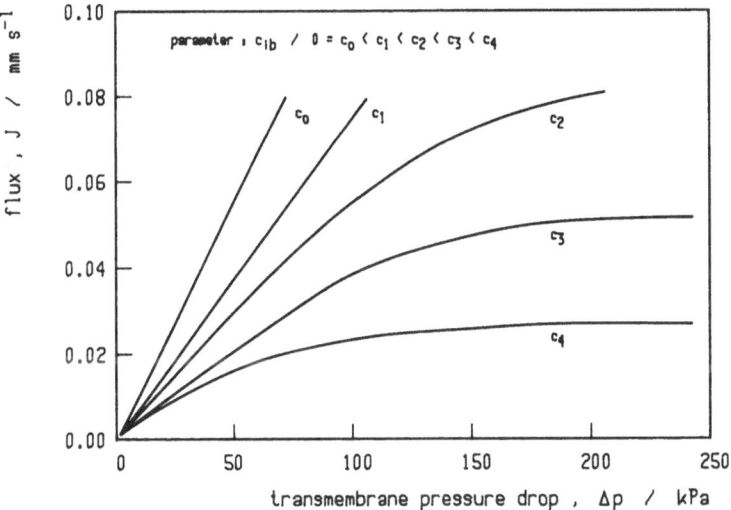

Fig. 2. Flux as a function of transmembrane pressure for different solute concentrations

Figure 3 shows the flux as a function of natural logarithm of the solute bulk phase concentration. The average transmembrane pressure is the parameter. For very low solute concentrations the flux is proportional to the pressure. With increasing concentration, the flux gradually slows especially at higher pressures. A linear relationship results when the secondary membrane starts to build up. For this situation the gel polarization model according to Eq. (10) can be applied:

$$J = k_d \ln (c_{is}) - k_d \ln (c_{ib}) \tag{15}$$

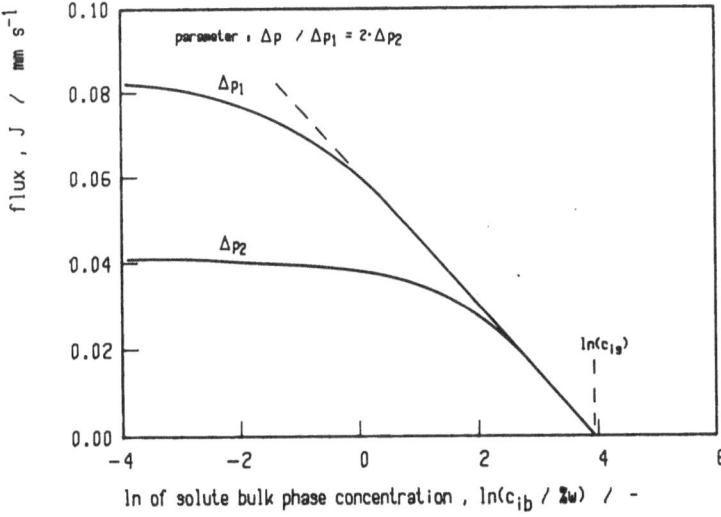

Fig. 3. Flux as a function of the natural logarithm of the solute bulk phase concentration

From this linear form of Eq. (10) it is evident that the mass-transfer coefficient (k_d) can be obtained from the slope and that extrapolation to $J = 0$ leads to the saturation concentration of the solute (c_{is}).

This method to determine the mass-transfer coefficient is widely used [6, 7, 17, 30, 31], because no knowledge of physical properties of the solution is needed. Since these are normally not known and the practical systems do not contain pure species, an experimental determination of k_d will always be necessary. The mass-transfer coefficient, determined in such a way, is only valid for the applied hydrodynamic conditions.

The hydrodynamics strongly influence the performance of ultrafiltration as shown in Fig. 4. The flux is given as a function of pressure at constant solute concentration with Reynolds number as parameter [6, 20, 21, 31]. At given module geometry, an increase of Reynolds number is proportional to an increase in the linear velocity of fluid along the membrane. High flow rates parallel to the membrane decrease the solute concentration at the membrane. The same result would be obtained by lowering the bulk phase concentration. Consequently it is not surprising that the family of curves in Fig. 4 is similar to that in Fig. 2.

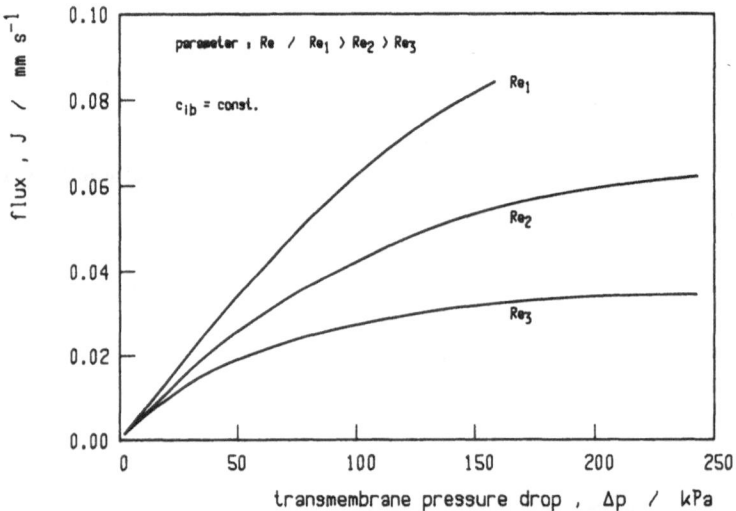

Fig. 4. Influence of the Reynolds number on flux as a function of transmembrane pressure

When the mass-transfer coefficient (k_d) is determined for different Reynolds numbers, as shown in Fig. 3, the correlation between the mass-transfer coefficient and the superficial velocity can be checked [6, 31, 32]. At laminar flow condition, Eqs. (12a), (12b) and for turbulent flow conditions Eqs. (12c), (12d) should describe this dependence. Together with Eq. (10), the flux can be related to the superficial velocity. Thus it is evident that a twofold increase of linear velocity should augment the flux by at least a factor of 1.26 in the laminar and 1.74 in the turbulent flow regime. The exponents in Eqs. (12a) — (12d) are obtained from double-logarithmic plots of flux versus either superficial velocity or Reynolds number [6, 7] as schematically shown in Fig. 5.

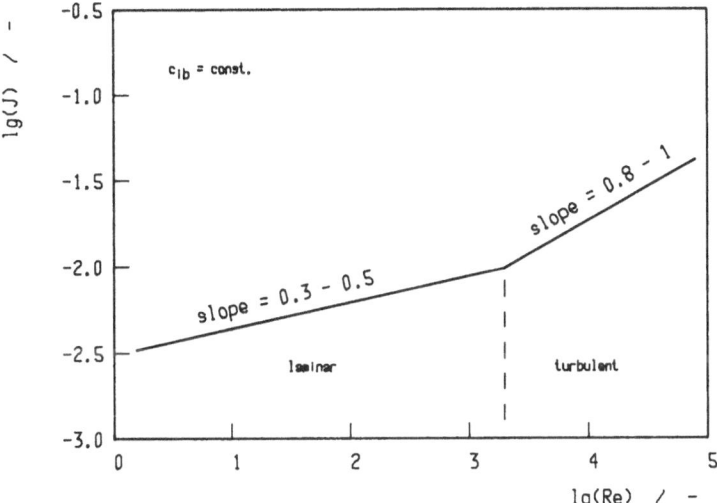

Fig. 5. Test for the validity of mass transfer correlations

Increasing the mean linear velocity will increase the pressure drop along the flow channel, while the average transmembrane pressure drop might remain constant. As illustrated in Fig. 6, this can lead to a decrease of the integral flux. It is assumed that for the lower Reynolds number (index 1) the transmembrane pressure at the entrance of the flow channel is so high that the solute saturation concentration is just reached at the membrane. Under such conditions, the flux hardly drops along the membrane. If the linear velocity is increased (index 2), the flux in the first half of the flow channel reaches the maximal possible value, but decreases steeply in the second half due to the steep slope of pressure drop along the flow channel. As can be seen from the hatched area in Fig. 6, the integral flux decreases at higher linear velocity. Obviously an optimum value of the Reynolds number exists as has been shown by Rauch [7]. When this value is exceeded, it is possible that a negative flux from the filtrate to the retentate occurs in the second half of the flow channel [33].

Although for concentration processes the aim is to attain quantitative retention,

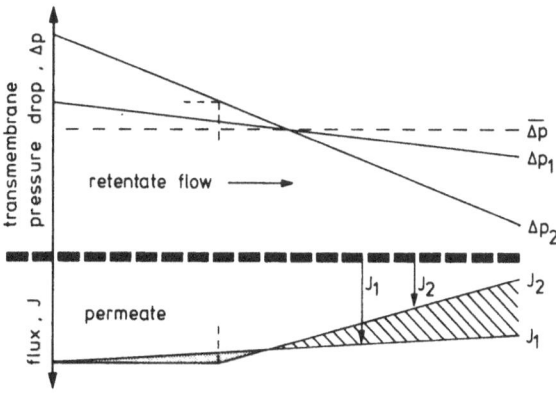

Fig. 6. Influence of the pressure drop along the membrane on the flux

this is not as critical as for other applications such as diafiltration or, in particular, membrane reactors. This results from the relatively short residence time of solute in the ultrafiltration unit during a concentration step. The integral loss of product is, therefore, rather low even if the retention is only found to be near unity.

Normally membranes are characterized according to their "cut-off". This value is given in molar mass and by convention means, that for a substance of that molar mass at least 90 % is retained. Since there is no agreement about a commonly accepted test method and standard test conditions, membranes with the same designation may vary considerably from one manufacturer to the other. Therefore, it is necessary to measure the retention under actual operating conditions.

For retention measurements an installation is recommended which allows control of the hydrodynamics independently of flux [34]. Only the apparent retention coefficient (R) can be obtained since it is not possible to determine the solute concentration directly at the membrane. While the intrinsic retention (R_m) is assumed to be a membrane property, the apparent retention (R) depends on the actual flow conditions. At low flux, the concentration polarization is negligible. Therefore, the concentration at the membrane can be assumed to be identical to that in the bulk phase. This implies that the intrinsic retention (R_m) will be obtained by extrapolation to zero flux. Since a higher flux increases the solute concentration at the membrane, the retention will decrease. This theoretical prediction has been confirmed by experiment [32, 35, 36], but it has to be kept in mind that most practical systems do not contain pure solute. Impurities of slightly higher molar mass than the desired substance enhance the retention [17, 32, 37]. Build-up of a gel layer can also improve retention owing to a lower 'pore size' of the secondary membrane. This occurs particularly at high solute concentrations [38].

As already discussed, an increase in Reynolds number normally has a positive effect on retention [30], but analogous to the flux, an optimum Reynolds number seems to exist [37], which is also due to the opposing influence of superficial velocity and pressure drop along the membrane (see discussion of Fig. 6).

Attempts have been made to use ultrafiltration for the fractionation of mixtures of macromolecular solutes. An ideal fractionation would require a membrane which retains only one species.

Sieve models for membranes have been developed [4] and a general correlation for their sieving curves has been proposed [39]. With a distinct sieving curve, the retention characteristic for different species can be calculated [40]. Butterworth and Wang [41] combined the concentration polarization theory (Eq. 8) with the definition of the retention coefficients to predict fractionation.

In practice, good separation selectivities can only be achieved for species with a great difference in their molar mass. This is due to a rather wide pore size distribution of most of the membranes and due to the build-up of a secondary membrane which changes the retention characteristic. The concentration polarization of the different species influences the retention of each other, so that each separation problem depends on the membrane, the hydrodynamics and the absolute and relative concentration of each species, respectively. Furthermore, it has to be considered that retention might be strongly influenced by pH or ionic strength in the case of charged solutes 'such as enzymes [41, 45]. Besides the single passage through a membrane [41], stage-wise operation has been proposed [42-44].

For a general discussion of fractionation, refer to Michaels [1], Butterworth and Wang [41] and Chapter 4.

Problems of shear deactivation during ultrafiltration of catalytically active species are discussed in the next chapter.

2.3 Membrane Reactors

The concept of ultrafiltration membrane reactors is based on the potential of semi-permeable membranes to retain the biocatalyst but not the products formed during the reaction. This implies that the molar mass of the products should be considerably lower than that of the catalyst. However, there is no limitation for the size of the substrate if it can be completely converted into products of low molar mass. One of the main advantages of membrane reactors is that the catalyst can be homogeneously distributed, thereby avoiding transport limitations. Therefore, this type of reactor should be potentially suitable for the depolymerization of natural macromolecules, reactions requiring pH-control and especially suitable for the operation with coenzyme-dependent enzyme systems.

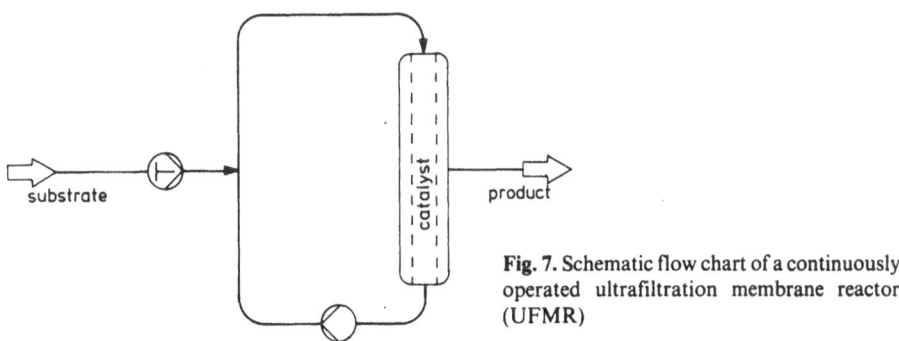

Fig. 7. Schematic flow chart of a continuously operated ultrafiltration membrane reactor (UFMR)

The concept of a continuously operated membrane reactor (UFMR) is shown in Fig. 7. Substrate solution is fed at constant flux to a recycle loop which contains, for example, a tubular ultrafiltration module. The biocatalyst is physically immobilized in the loop, since it is retained by the membrane. Products of low molar mass leave the system with the filtrate. If the catalyst activity drops due to deactivation, additional biocatalyst can be added to the reactor in order to keep the substrate conversion at the desired level. For membrane reactor operation, a very high biocatalyst retention must be achieved. When the mass balance for the biocatalyst (e.g. enzyme) is formulated for the case of incomplete retention, the result is:

$$\frac{dE}{dt} = - J \frac{A_m}{V_R} E_f = - J a_m E_f \qquad (16)$$

where

A_m	m^2	membrane area
V_R	m^3	reactor volume

a_m $m^2\,m^{-3}$ volume specific membrane area
E $kg\,m^{-3}$ active enzyme concentration (assuming homogeneous distribution)
E_f $kg\,m^{-3}$ enzyme concentration in the filtrate

The product of flux and volume specific area is identical to the reciprocal residence time:

$$\frac{dE}{dt} = -\frac{1}{\bar{t}}E_f \tag{17}$$

where

$\bar{t}$ s   mean residence time

With $E_f = (1 - R)\,E$ this leads to:

$$\frac{dE}{dt} = -\frac{1}{\bar{t}}(1 - R)\,E \tag{18}$$

This equation can be integrated taking into account the catalyst concentration at time zero:

$$\frac{E}{E_o} = e^{-\frac{1-R}{\bar{t}}t} \tag{19}$$

where

E_o $kg\,m^{-3}$ initial catalyst (enzyme) concentration
t s   time of operation

The decay of activity depends on the retention coefficient and the residence time. A deactivation constant can be derived according to its definition:

$$k_{de} = \frac{1}{E}\frac{dE}{dt} = \frac{1 - R}{\bar{t}} \tag{20}$$

where

$k_{de}\ s^{-1}$ deactivation rate constant

In Fig. 8, the residual activity is given as a function of dimensionless time $(t/\bar{t})$ for different retention coefficients. As can be seen from Fig. 8, the activity loss can be quite important even for a rather high catalyst retention. On the basis of a retention coefficient of 99 % and a residence time of 1 h, 21.3 % of the catalyst will be lost after one day of operation.

The apparent retention coefficient (R) can be obtained graphically from the slope of a plot of the actual enzyme concentration versus operating time:

$$\ln E = \ln E_o - \frac{1 - R}{\bar{t}}t \tag{21}$$

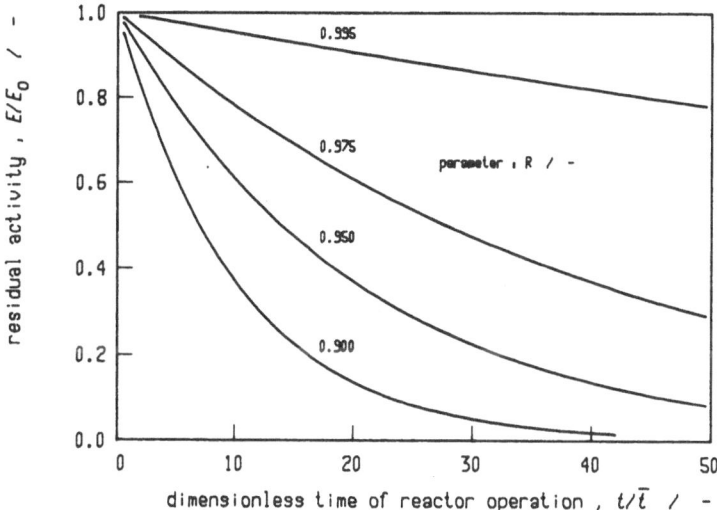

Fig. 8. Loss of activity during the operation of a UFMR due to non-quantitative retention of the catalyst

It should be kept in mind that the apparent retention depends on the hydrodynamics (Sect. 2.2). Since membrane reactors should be operated with membranes of quantitative catalyst retention, each ultrafiltration unit must be examined for even small deviations from this requirement.

As for other catalytic reactors, the substrate conversion in membrane reactors depends on the residence time and catalyst concentration. The residence time is theoretically limited by the hydraulic permeability of the ultrafiltration unit. The required catalyst concentration is normally low in comparison to the solute concentration in separation processes. Nevertheless, the pressure drop across the membrane increases sharply during the start up procedure due to adsorption of catalyst on the membrane [46]. This influences the distribution of the biocatalyst and the average transmembrane pressure.

Adsorption can also influence the catalyst retention. Normally the retention is improved by partial clogging of the pores by penetrating particles [40]. For further information on retention Sect. 2.2 should be consulted.

For membrane reactor application not only quantitative retention of the catalyst is required, but the catalyst also has to exhibit appropriate stability. Biocatalysts are affected by chemicals, heat and other physical conditions. Since the chemical and thermal effects are, in principle, no different in membrane reactors than in other reactors, only the physical effects will be considered here. As already mentioned, surfaces including that of the membrane can cause catalyst deactivation due to adsorption or incompatibility. At very low catalyst concentrations, a quite detectable fraction of the catalyst can adsorb on the membrane. Taking into account the build-up of a secondary membrane at high catalyst concentrations, an optimum between these limits should exist.

In order to avoid concentration polarization, high linear velocities parallel to the membrane have to be applied. In consequence, high shear rates will occur especially near the membrane. Deactivation of enzymes and denaturation of proteins due to shear stress has been studied in model systems such as capillary tubes, visco-meters [47-51], and rotating cylinders [52]. Similar effects have also been found during ultrafiltration [53, 54] and pumping [51, 55]. It has been reported that shearing of enzymes may alter their kinetics [49, 56-58]. For a general survey of this subject see Charm and Wong [59]. Although there are several authors who attribute the cause for deactivation to influences other than shearing [56, 57, 60, 61], the effect may vary considerably from one species to another, but cannot be completely neglected. Thomas and Dunnill [60] have gathered a rather complete collection of effects which are either known or supposed to adversely affect biocatalysts.

For laminar flow, several attempts have been made to modelling this phenomenon. In tubular flow channels, a parabolic velocity profile develops during laminar flow:

$$u_r = (r_t^2 - r^2) \frac{1}{4\eta} \frac{\Delta p}{l_t} \tag{22}$$

where

u_r $m\,s^{-1}$ linear velocity at radial position r
r m   radial position $(0 \leqq r \leqq r_t)$
r_t m   tube radius

The shear rate is proportional to the velocity gradient:

$$\gamma_r = -\left(\frac{du}{dr}\right)_r = \frac{1}{2\eta} \frac{\Delta p}{l_t} r \tag{23}$$

where

γ_r s^{-1} shear rate (absolute value) at radial position r.

The shear rate is linearly related to the radius and attains its maximum value at the wall $(r = r_t)$.

It is assumed that the shear rate is responsible for the destabilisation of biocatalysts. However, in practice, only the average deactivation in a flow channel can be measured. For correlating the rate constant of deactivation and the shear rate, the local shear rate has to be integrated over the cross section and has to be weighted with the local mass flow. For this purpose, the use of the average shear rate over the channel cross section has been suggested [47]:

$$\bar{\gamma}_r = \int_0^{r_t} \left(\frac{1}{2\eta} \frac{\Delta p}{l_t} r\right) \frac{2\pi r\, dr}{\pi r_t^2} \tag{24}$$

where

$\bar{\gamma}_r$ s^{-1} av. shear rate $\left(= \dfrac{1}{3\eta} \dfrac{\Delta p}{l_t} r_t\right)$

It has also been suggested that the variation of residence time for particles flowing on different stream lines should be taken into account and averaged over the cross section [47]. Since the residence time at the wall is infinite, such an averaging is not possible. The analytical solution given by Charm and Wong [47, 59] is invalid for the assumptions made ('mass average shear'). A closer analysis shows that the mean hydrodynamic residence time has been applied, what leads to a rather over-simplified solution.

This concept is misleading because the deactivation constant may not be proportional to the shear rate. The average hydrodynamic residence time does not take into account that quite different amounts of biocatalyst 'see' different shear rates as a function of their radial position. If the dimensionless product of average residence time and average shear rate is used as the characterizing parameter, the implication is that doubling the shear stress and halving the residence time will have no effect. This is certainly not the case because shear deactivation is, according to Charm and Wong [59], an activated process just like thermal deactivation.

$$k_{de} = k_{de}^o \, e^{-\left(\frac{E_a/\Delta V}{\tau}\right)} \tag{25}$$

where

k_{de}	s^{-1}	deactivation constant (shear only)
E_a	$J \cdot mol^{-1}$	activation energy
ΔV	$m^3 \, mol^{-1}$	change of molar volume
τ	Pa	shear stress $(= \eta \gamma)$

Since the shear stress in tubular geometries is anisotropic, in contrast to the temperature, Eq. (25) can only be valid for a particular streamline:

$$k_{de}^r = k_{de}^o \, e^{-\left(\frac{E_a/\Delta V}{\frac{\Delta p}{l_t} \frac{r}{2}}\right)} \tag{26}$$

where

k_{de}^r s^{-1} deactivation rate constant at radial position r

This function shows that the shear deactivation increases in a non-linear manner from the center of the tube to the wall. To obtain the mass average shear deactivation, each value of k_{de}^r has to be weighted with the corresponding differential mass flow at the same radial position. This weight function describes the mass flow for each entire streamline relative to the mean mass flow:

$$\frac{1}{\bar{m}} \frac{d\dot{m}}{dr} = \frac{4}{r_t} \left(\frac{r}{r_t} - \frac{r^3}{r_t^3}\right) \tag{27}$$

with

$$\frac{d\dot{m}}{\bar{m}} = \frac{u_r E}{\bar{u} E} \frac{dA}{A_t} = \frac{u_r}{\bar{u}} \frac{2r}{r_t^2} dr$$

where

A_t m^2 cross section (tube)

$\bar{u}$ $m\,s^{-1}$ mean linear velocity $\left(= \dfrac{1}{8\eta}\dfrac{\Delta p}{l_t} r_t^2 \right)$

E $kg\,m^{-3}$ concentration of enzyme (assumed to be independent of radial position)

The corresponding function is given for $r_t = 1$ mm in Fig. 9.

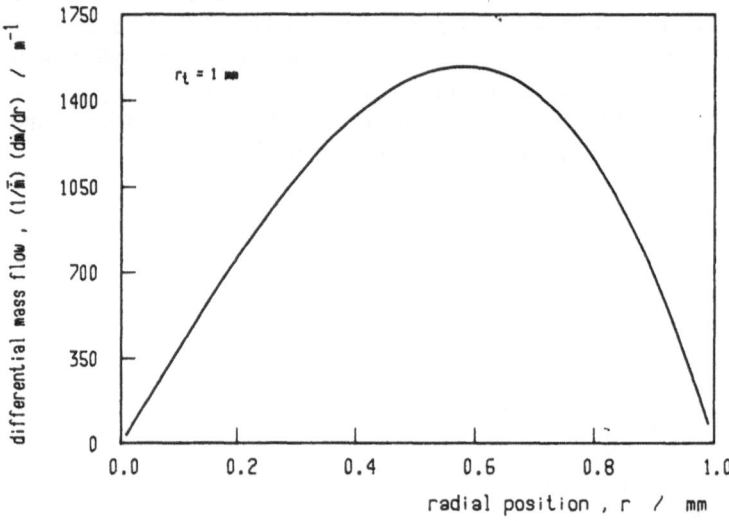

Fig. 9. The distribution of mass flow on streamlines at different radial position for laminar flow in tubes

The average deactivation constant can be obtained by combining Eqs. (26) and (27):

$$\overline{k_{de}} = \int_0^{r_t} k_{de}^r \frac{1}{\dot{\bar{m}}} \frac{d\dot{m}}{dr}\, dr \tag{28}$$

or in detail

$$\overline{k_{de}} = \int_0^{r_t} k_{de}^o\, e^{-\left(\frac{E_a/\Delta V}{\frac{\Delta p}{l_t}\frac{r}{2}}\right)} \frac{4}{r_t}\left(\frac{r}{r_t} - \frac{r^3}{r_t^3}\right) dr \tag{29}$$

The corresponding function is shown schematically in Fig. 10. The area under the curve represents the mass average deactivation constant. From this figure, it can be seen that no contribution is expected to come from the center of the tube due to

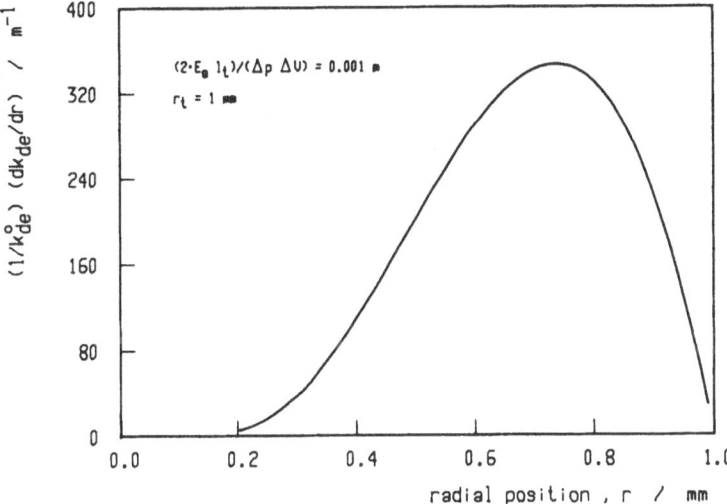

Fig. 10. The radial distribution of shear deactivation for laminar flow in tubes

absence of shear. In addition, no contribution should come from positions near the wall, in spite of the highest shear rates, due to the negligible differential mass flow. The developed theory on shear deactivation (Eq. (29)) shows that the interrelation of the variables is much more complex than has been suggested before. This model should only be applied in the laminar flow range and under conditions where the catalyst is equally distributed over the cross section. This theory might be helpful in understanding and interpreting model experiments on shear deactivation.

3 Process Design

The design of ultrafiltration processes depends on the mode of separation. Three distinct modes have to be considered: concentration, diafiltration and fractionation. During concentration the solvent has to be removed, whilst during diafiltration low molar mass solutes are replaced by pure solvent. Fractionation can resemble either of these two processes.

Continuous biocatalytic reactions in membrane reactors resemble dialfiltration, since low molar mass product is replaced by substrate solution.

3.1 Design and Operation of Separation Units

Concentration

Concentration processes are carried out in batch or continuous mode of operation. The main design objective, besides quantitative retention, is an optimum use of the ultrafiltration unit for rapid achievement of high solute concentrations with minimum energy input and minimum catalyst deactivation.

Figure 11 shows a flow scheme for a multi-purpose ultrafiltration unit. The inner recycle loop contains the ultrafiltration module and the recirculation pump P_2. It is connected to a storage tank and to the pump P_1. A four way distributor allows for the different modes of operation illustrated on the flow chart.

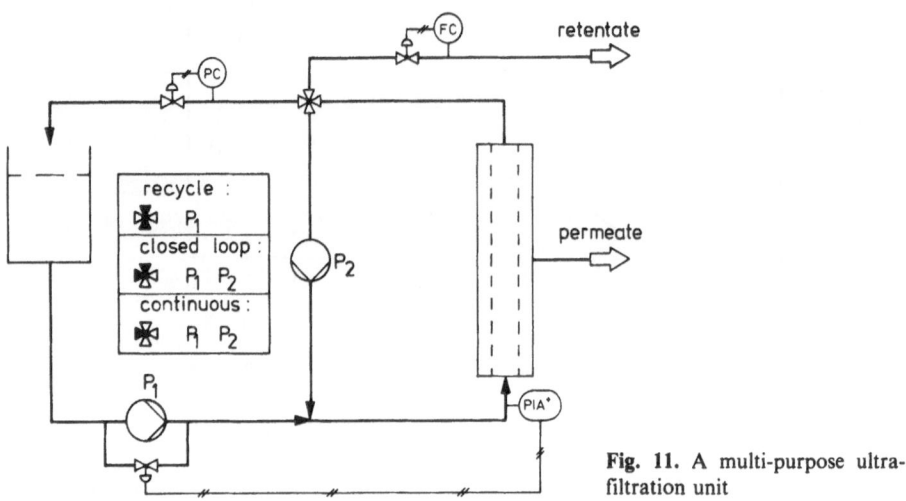

Fig. 11. A multi-purpose ultra-filtration unit

Under batch recycle operation, the pump P_2 is turned off, and the entire solution is recirculated by means of pump P_1. In order to obtain and maintain an appropriate transmembrane pressure, a control valve is installed in the return line to the storage tank. Filtrate is continuously withdrawn and the solute concentration increases simultaneously in the whole system.

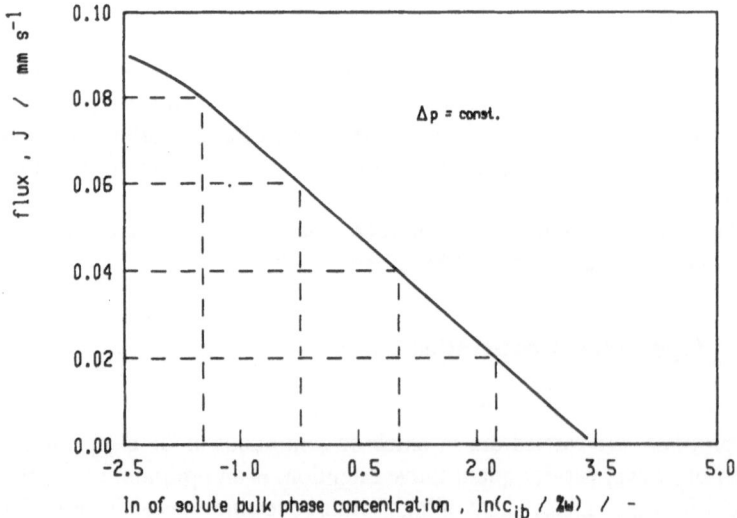

Fig. 12. Schematic representation of the theoretical background of concentration processes

In a slightly different mode of batch operation, named closed loop, the concentrate is only recycled in the inner loop containing pump P_2. With this mode of operation both pumps have to be operated and the catalyst concentration at the membrane increases more rapidly than with the recycle mode. The advantage of this mode of operation is that the hydrodynamics at the membrane can be adjusted independently of the feed rate and the control valve which might cause shear deactivation can be omitted.

Continuous operation requires a bleed stream of retentate. This is regulated by a flow controller in order to achieve the desired solute concentration.

To illustrate the differences between the discussed modes of operation, the achievable flux as a function of the actual bulk phase concentration is shown in Fig. 12 for a constant average transmembrane pressure. From this the conclusion has to be drawn that the optimum, with respect to operating time, occurs when the solute concentration is kept low. On this basis, recycle operation should be selected. The objective of keeping the solute concentration as low as possible, is not achieved unless the concentrate is rediluted with the entire remaining solution. With respect to membrane utilisation, continuous operation is even worse, since the ultrafiltration unit will operate at the lowest flux, because the solute concentration in the module corresponds to the desired exit concentration.

If continuous operation is essential, ultrafiltration modules should be arranged in series as shown in Fig. 13. For a given feed rate, the bleed stream of the concentrate is kept constant by means of a regulating valve at the end of the last stage. If the main pressure drop occurs across the membrane, the entire system will operate at nearly constant average transmembrane pressure. According to the relationship between the over-all solvent flux and the flow rate of concentrate, the flux is quite different for each stage. This is due to increasing solute concentrations from stage to stage. Theoretically, a very long tubular ultrafiltration device would be optimal, since the flow of the filtrate would lead to a gradual increase of solute concentration along the flow channel.

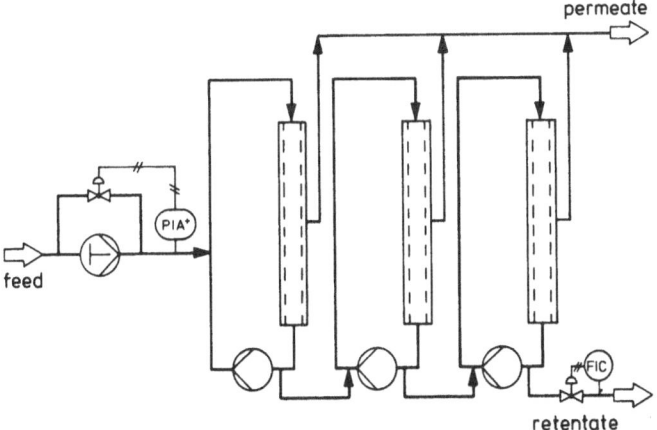

Fig. 13. Flow chart of an ultrafiltration cascase for continuous concentration

The design objective for batch concentration using recycle operation is to determine the required membrane area for concentrating a given volume in a certain operating time. The solute concentration increases from the initial value to the final value, corresponding to the change of volume:

$$c_{ib} = c_{io} \frac{V_o}{V}$$

(30)

where

c_{io} kg m^{-3} initial solute concentration
V_o m^3 initial volume
V m^3 actual volume

If it is assumed that the saturation concentration of solute is attained at the membrane, the gel polarization model (Eq. (10)) can be applied:

$$J = -\frac{1}{A_m} \frac{dV}{dt} = k_d \ln \frac{c_{is}}{c_{ib}}$$

(31)

where

A_m m^2 membrane area

Combining Eqs. (30) and (31) and rearranging leads to:

$$-\int_{V_o}^{V_e} \frac{dV}{\ln \left(\dfrac{c_{is}}{c_{io}} \dfrac{V}{V_o} \right)} = k_d A_m \int_0^{t_e} dt$$

(32)

where

V_e m^3 final volume
t_e s   required operating time

This equation can be solved by partial integration leading to a series. It can also be expressed in the form:

$$-\frac{dV}{dt} = A_m \left(J_o - k_d \ln \frac{V_o}{V} \right)$$

(33)

where

$$J_o \quad \text{m}^3 \text{ m}^{-2} \text{ s}^{-1} \quad \text{initial flux} \quad \left(= k_d \ln \frac{c_{is}}{c_{io}} \right)$$

This differential equation can be integrated numerically to obtain the required operating time (t_e). This time depends not only on the volume ratio (volume reduction) but also on the initial flux.

For closed loop operation, a similar equation can be derived which depends on the initial conditions. Assuming that the loop is initially filled with the solution and the initial volume in the feed tank is referred to this state, the mass balance for the solute is:

$$c_{ib} = c_{io}\left(1 + \frac{V_o - V}{V_S}\right) \tag{34}$$

where

V_S m^3 volume of the loop (separation unit)

Combining this with the gel polarization model (Eq. (10)) leads to:

$$-\frac{dV}{dt} = A_m\left[J_o - k_d \ln\left(1 + \frac{V_o - V}{V_S}\right)\right] \tag{35}$$

The shortest operating time will be obtained at the maximum admissible average transmembrane pressure and optimum hydrodynamic conditions. 'Optimum' implies that a suitable Reynolds number has to be chosen not only to achieve a high flux, but also to avoid high pressure drop along the flow channel and shear deactivation of sensitive biocatalysts.

For continuous concentration, the mass balance for the solute is:

$$\dot{V}_f = \dot{V}_o\left(1 - \frac{c_{io}}{c_{ie}}\right) \tag{36}$$

where

$\dot{V}_f$	m^3 s^{-1}	volumetric flow rate of filtrate
$\dot{V}_o$	m^3 s^{-1}	volumetric flow rate of feed
c_{io}	kg m^{-3}	solute concentration in the feed stream
c_{ie}	kg m^{-3}	solute concentration in the retentate (exit stream)

The volumetric flow rate of permeate and the flux are correlated by means of the membrane area (A_m):

$$J = \frac{\dot{V}_f}{A_m} = k_d \ln\frac{c_{is}}{c_{ie}} \tag{37}$$

This equation can be solved for the membrane area:

$$A_m = \frac{\dot{V}_o\left(1 - \dfrac{c_{io}}{c_{ie}}\right)}{k_d \ln\dfrac{c_{is}}{c_{ie}}} \tag{38}$$

The required membrane area can be calculated for a given throughput and concentration ratio. Since the entire system operates at the retentate concentration (c_{ie}), the lowest possible flux is obtained.

The required membrane area depends obviously on the mass-transfer coefficient. Even when the gel polarization model can be applied, some dependence on the type of membrane will persist. The film theory can normally be applied only for pure systems. For complex mixtures, the physical properties are unknown and the effective mass-transfer coefficient has to be evaluated by experiment (refer to Chapter 2.2). In most practical cases, it will be more convenient to evaluate the mass-transfer coefficient as a function of superficial velocity experimentally, than to determine the diffusion coefficient of the solute with adequate accuracy. Nevertheless, the film theory can serve as a first approximation for the design of ultrafiltration units.

For continuous concentration processes, the concentration ratio achievable in single stage ultrafiltration has to be compared to that obtained in cascades:

$$\frac{c_{i(n-1)}}{c_{in}} = \frac{\dot{V}_{n-1}}{\dot{V}_{n-1} - A_m k_d \ln \dfrac{c_{is}}{c_{in}}} \tag{39}$$

and

$$\dot{V}_n = \dot{V}_{n-1} - \dot{V}_{f(n-1)} = \dot{V}_{n-1} \frac{c_{i(n-1)}}{c_{in}}$$

where

n — number of stage

A numerical comparison is given in Fig. 14. The saturation concentration is assumed to be $c_{is} = 20$. The other variables in Eq. (39) are equated to unity. Whilst the single stage unit with three parallel modules yields a final concentration of 73.4% of the saturation concentration, the three stage cascade yields a final concentration of 99.7%. This indicates that staged installations require much less membrane area than single units, for the same degree of concentration or volume reduction.

Diafiltration

For diafiltration, it is assumed that the biocatalyst is retained quantitatively and that low molar mass molecules (for convenience, salts) pass through the membrane without hindrance. The main objective is to remove these substances with a reasonable membrane area in a given operating time with the consumption of as little additional pure solvent (eluant) as possible. With respect to flux, there is, in principle, no difference between the process of concentration and diafiltration. Already, during the concentration step, a considerable amount of salt is removed together with solvent. Therefore, dialfiltration is normally undertaken after a concentration step.

The multi-purpose ultrafiltration unit shown in Fig. 11 can be used. When a reasonable catalyst concentration in the retentate is reached during recycle operation,

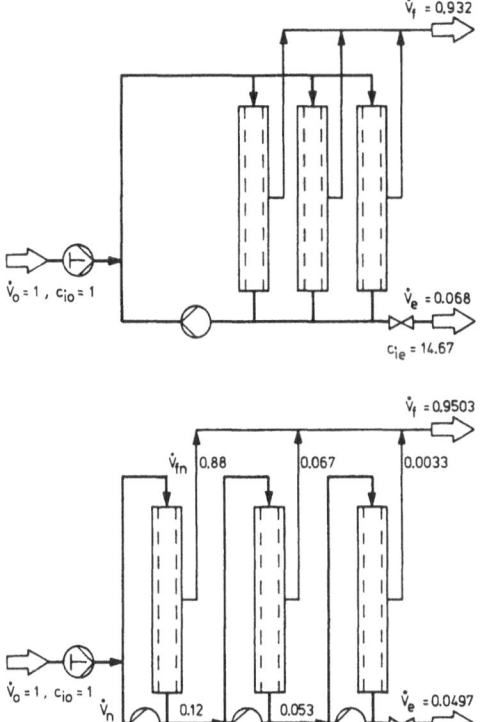

Fig. 14. Comparison of a normal and staged ultrafiltration for continuous concentration

the operating mode is changed to closed loop. Eluant is then continuously added by means of the pump P_1, while the corresponding amount of salt solution is forced across the membrane. This wash-out process depends on the spent volume of eluant:

$$\frac{c_i}{c_{io}} = e^{-\frac{V_w}{V_o}}$$ (40)

where

c_i	$kg\ m^{-3}$	actual salt concentration
c_{io}	$kg\ m^{-3}$	initial salt concentration
V_w	m^3	volume of added eluant
V_o	m^3	initial retentate volume

Another possibility is to maintain the recycle operation for diafiltration. The storage tank is filled with eluant and the solution is re-concentrated. Assuming that the initial and final volumes remain equal in each dilution/concentration-cycle, the mass balance is:

$$\frac{c_{in}}{c_{io}} = \left(\frac{1}{1 + \frac{V_w}{V_o}\frac{1}{n}}\right)^n$$ (41)

where

n   —   number of operating cycles
V_w m³   entire volume of added eluant

In Table 2, the final relative concentration of salt in the retentate is given for different numbers of cycles. The second parameter is the volume ratio of the total spent eluant, distributed over different numbers of cycles, to the initial catalyst solution (V_w/V_o).

Table 2. The reduction of salt concentration during diafiltration by repetitive dilution/concentration cycles and closed loop operation ($n = \infty$). The total eluant volume (V_w) is distributed over a varied number of cycles

n	1	2	3	10	∞
V_w/V_o	c_i/c_{io}				c_i/c_{io}
1	0.5	0.444	0.422	0.386	0.368
2	0.33	0.25	0.216	0.162	0.135
3	0.25	0.16	0.125	0.073	0.049
10	0.091	0.028	0.012	0.001	4.5×10^{-5}

Continuous diafiltration is advantageously carried out in staged plants. Such an installation is shown in Fig. 15. The concentrated solution is fed at constant rate whilst the purified catalyst leaves the installation at the same rate and usually at the same concentration. This is achieved by means of a control valve connected to a flow meter. Eluant is fed in cross-stream to each stage. Since the retained solute is equally distributed over the stages, the flux in each stage should be the same. Therefore, the feed rate of eluant can be kept constant by means of a pump of adjustable speed

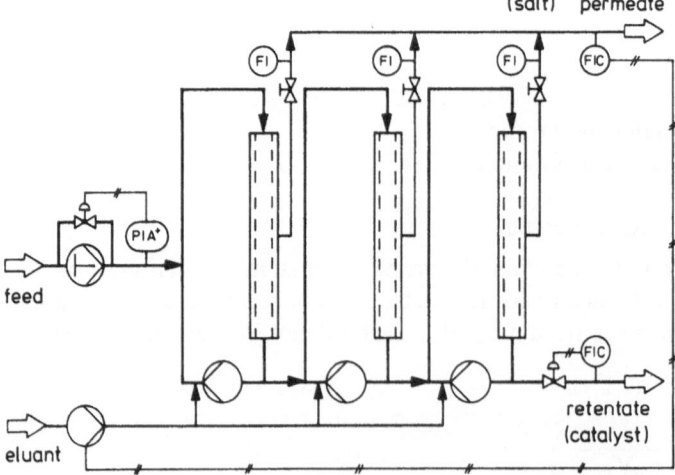

Fig. 15. Flow chart of an ultrafiltration cascade for continuous diafiltration

which is controlled according to the actual flow rate of permeate. This pump can be either a reciprocating or, simply, a pressurizing one. To obtain the mass balance of a diafiltration cascade, Eq. (41) can be used. Here, n is the number of stages and the volume ratio has to be replaced by the feed rate ratio ($\dot{V}_w/\dot{V}_o$). An infinite number of stages leads to an idealized tubular cross flow system which would be the optimum configuration, although essentially hypothetical. In Fig. 16, the required eluant volume, referred to that required for a single stage diafiltration unit, is given as a function of the desired relative residual salt concentration. The number of stages in the diafiltration cascade is given as the parameter. The volume ratio also indicates the ratio of the required membrane areas since the membrane area which has to be installed is proportional to the total volume of eluant.

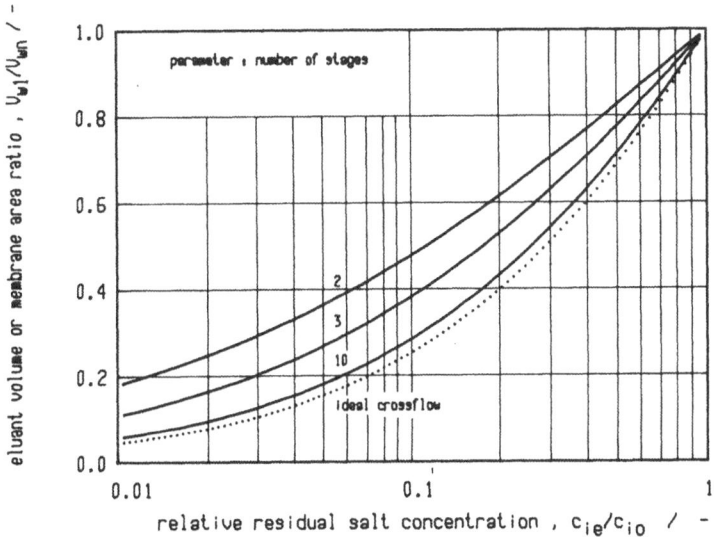

Fig. 16. Comparison of the required eluant volume (membrane area) of a single ultrafiltration unit with cascades for continuous diafiltration

It has been pointed out that diafiltration is theoretically best achieved by infinitesimal dilution/concentration steps. For large-scale plants, staged operation can be applied, similar to concentration processes. In contrast to the latter, the retained solute is equally distributed over the diafiltration plant. This means that the variation of flux as a function of solute concentration can be neglected when the different modes of operation are compared. It turns out that the concept of minimizing the required eluant volume will also minimize, as a first approximation, the required membrane area. On the other hand, the design equations given for concentration processes must be applied to calculate the membrane area for the desired operating time.

3.2 Design and Operation of Membrane Reactors

Since questions like the applicability of membrane reactors and catalyst retention, as well as shear deactivation, have been discussed in Chapter 2.3, this chapter will deal exclusively with the different operating modes for membrane reactors. For convenience, it will be assumed that the actual catalyst is an enzyme which is quantitatively retained by the ultrafiltration membrane. Deactivation is considered as an over-all effect and not differentiated according to each particular cause. Only flat membranes and tubular modules will be considered. For clarity, an ultrafiltration membrane reactor (UFMR) will be defined as a reaction system, in which the catalyst is by preference homogeneously distributed. This is necessary for discrimination of this type of reactor from membrane reactors, which employ catalysts gelified or bound on the membrane and from porous wall tubular reactors (PWTR).

The required reaction engineering fundamentals must be given before an optimization strategy for continuously operated membrane reactors can be discussed. The latter combines reaction engineering concepts with the theory of ultrafiltration.

A discussion of reactors cannot be accomplished without taking into account the kinetics of the reaction systems in question. Thus, homogeneous enzymatic catalysis must be considered. To show the influence of kinetics on reactor design, a representative example should be used. Since most of the enzymatic reactions of industrial interest exhibit either simple or complex product inhibition (e.g. glucose isomerase, amylases, cellulases, acylases, lactases), sometimes combined with a reverse reaction (e.g. glucose isomerase, acylase), the kinetic model of Michaelis-Menten for competitive product inhibition is taken. For a single substrate reaction the basic expression is:

$$r_v = S_o \frac{dX}{dt} = k_2 E \frac{S_o(1-X)}{K_m \left(1 + \frac{S_o X}{K_{ic}}\right) + S_o(1-X)} \tag{42}$$

where

r_v	mol m^{-3} s^{-1}	volume specific reaction rate (absolute value)
S_o	mol m^{-3}	initial substrate concentration
X	—	degree of substrate conversion
k_2	mol kg^{-1} s^{-1}	reaction rate constant
E	kg m^{-3}	enzyme concentration (active)
K_m	mol m^{-3}	constant of Michaelis-Menten
K_{ic}	mol m^{-3}	constant of competitive product inhibition

Thus, the reaction rate is either a function of both the initial substrate concentration and the conversion, or of both the actual substrate concentration ($S_o(1-X)$) and the actual product concentration ($S_o X$). Neglecting inhibition ($K_{ic} = \infty$), the reaction rate is determined by the maximum enzyme activity, expressed by the rate constant (k_2) and the Michaelis-constant (K_m). A small K_m value signifies a zero order reaction up to high substrate conversions, whilst at high K_m values a nearly first order reaction is obtained which is characterized by a linear decline of reaction rate with conversion. If competitive product inhibition is taken into account, the decline can be even steeper than first order, as shown in Fig. 17. The normalized reaction rate is:

$$\frac{r_v}{V_{max}} = \frac{r_v}{k_2 E} \tag{43}$$

where

V_{max}	mol m^{-3} s^{-1}	maximum reaction rate

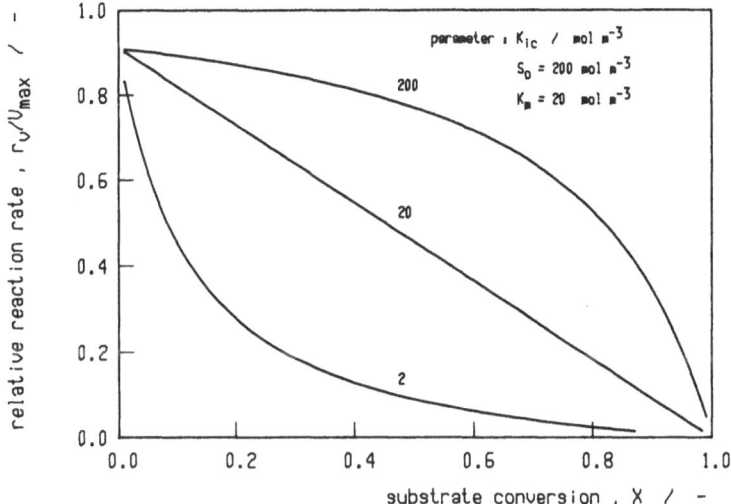

Fig. 17. The normalized reaction rate as a function of substrate conversion for different extents of product inhibition

This results for small inhibition constants since the K_m value is amplified during the reaction by addition of $(K_m/K_{ic}) S_o X$, a function of product concentration. Therefore, the degree of competitive product inhibition is not only determined by the inhibition constant alone, but to a similar extent by K_m.

To deduce strategies for the design of membrane reactors, the kinetics have to be combined with the particular reactor characteristics. This can be discussed in terms of the two extreme cases: the ideal continuous stirred tank reactor (CSTR) and the ideal plug flow tubular reactor (PFTR). The ideal CSTR is characterized by complete backmixing, which means homogeneous distribution of both catalyst and reactants. In other words, this type of reactor works at outlet concentrations with respect to both substrate and product. Assuming that the CSTR is fed with pure substrate solution, the mass balance is:

$$S_o \frac{X}{\bar{t}} = r_v = E r_m \tag{44}$$

where

$\bar{t}$ s   mean residence time
r_m mol kg^{-1} s^{-1} (catalyst-) mass specific reaction rate (absolute value)

This can be rearranged to yield the required (catalyst-) mass referred space time:

$$\tau_m = E \bar{t} = \frac{1}{r_m} S_o X \tag{45}$$

and in analogy

$$\tau_m = \frac{m_E}{V_R}\,\bar{t} = \frac{m_E}{\dot{V}_R} = \frac{1}{r_m}\,S_o X \tag{46}$$

where

τ_m	kg s m^{-3}	(catalyst-) mass referred space time
m_E	kg	mass of enzyme (active)
V_R	m^3	reactor volume
$\dot{V}_R$	m^3 s^{-1}	volumetric feed rate

Using Eqs. (45) and (46) in conjunction with the particular kinetics (r_m) (e.g. Eq. (42)), the operating variables, either $\bar{t}$ and E or m_E and $\dot{V}_R$, can be calculated for given initial substrate concentrations and the required conversion. For steady state kinetics the enzyme concentration and the mean residence time are of equal influence. When a particular conversion is required at a specified initial substrate concentration, the space time τ_m has to be adjusted to a distinct value by regulating either E and $\bar{t}$ or m_E and $\dot{V}_R$ to any combination which yields the required value for τ_m. τ_m can be readily obtained from the mass balance. The expression of substrate conversion as a function of τ_m normally yields quadratic equations which can be analytically solved.

The main design parameter will be the productivity per unit mass of enzyme:

$$\bar{r}_m = \frac{S_o X}{E\bar{t}} = r_m \tag{47}$$

where

$\bar{r}_m$	mol kg^{-1} s^{-1}	(catalyst-) mass specific mean reaction rate ($=$ productivity per unit mass of enzyme)

This is, for a CSTR, equal to the kinetically determined reaction rate at the desired S_o and X. Considering the kinetics shown in Fig. 17, the productivity can be rather low at high degrees of conversion.

The reaction in an ideal plug flow tubular reactor (PFTR) takes place differentially. Thus, the substrate concentration decreases along the length of the reactor. When this is expressed by a transformation to the time coordinate, the balance for a PFTR is equal to that of an ideal stirred batch reactor (SBR):

$$\frac{S_o}{E}\frac{dX}{dt} = r_m \tag{48}$$

This can be rearranged and integrated to yield the required space time τ_m as a function of S_o and X:

$$\tau_m = E\bar{t} = S_o \int_0^X \frac{1}{r_m}\,dX \tag{49}$$

This expression can be solved analytically for steady state kinetics. The conversion, as a function of τ_m, cannot be expressed analytically and must be solved by iteration, since the integral in Eq. (49) leads to expressions containing the natural logarithm of conversion.

The productivity of a PFTR or SBR is:

$$\bar{r}_m = \frac{S_o X}{E \bar{t}} = \frac{X}{\displaystyle\int_0^X \frac{1}{r_m} dX} \tag{50}$$

The PFTR gives a higher productivity than the CSTR because it works at higher substrate concentrations, from S_o until $S_o(1 - X)$ is attained at the outlet. The productivity relationship between a CSTR and a PFTR is found to be:

$$\frac{\bar{r}_m (CSTR)}{\bar{r}_m (PFTR)} = \frac{r_m}{X} \int_0^X \frac{1}{r_m} dX \tag{51}$$

This relation is obviously unity at zero conversion and tends to zero for quantitative conversion. This relation is shown in Fig. 18 for the same kinetics and conditions applied in Fig. 17. When highly inhibited reactions are carried out or nearly quantitative conversion is needed, the catalyst is much more efficiently used in a PFTR than in a CSTR.

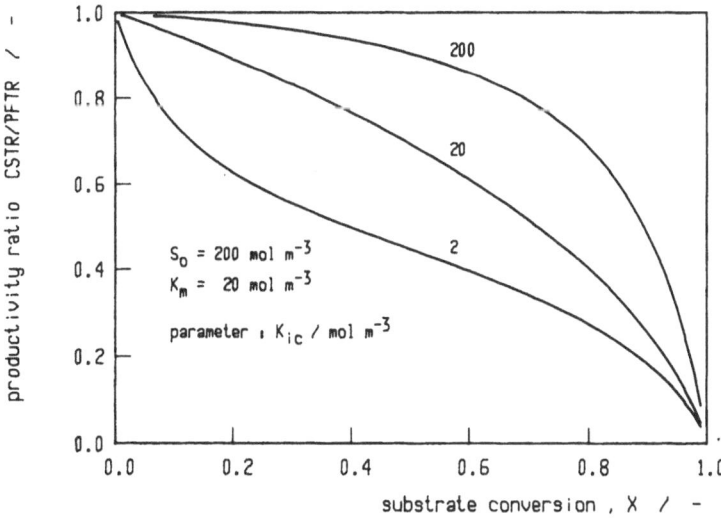

Fig. 18. Comparison of the productivity of an ideal continuous stirred tank reactor (CSTR) with a plug flow tubular reactor as a function of substrate conversion for different extents of product inhibition

For the discussion of membrane reactors, this leads to the conclusion that backmixing should be avoided, since it tends to dilute the actual substrate concentration. The simplest way to achieve this is to use a mode of operation described as 'repetitive batch', i.e., the reaction being carried out in a stirred batch reactor (SBR). After obtaining the desired conversion, the reaction mixture is subjected to ultrafiltration for the recovery of the catalyst. This can be achieved by any of the techniques described in the previous chapter, and especially by the economical recycle operation. The concentrated enzyme solution is flushed back to the reactor by passing new substrate solution through the ultrafiltration plant while the permeate outlet is closed.

Since the reaction and the ultrafiltration are uncoupled, this process is easily designed. For the reaction step, only the relationship between the mass referred space time and conversion must be known according to Eq. (49), for the desired initial substrate concentration. In combination with experimental data on ultrafiltration of the catalyst in the form of flux as a function of enzyme concentration, a convenient enzyme concentration can be determined to assure a practical cycle time for operation. Thus taking into account Eq. (49) for the reaction, a numerical solution of Eq. (33) for ultrafiltration and the over-all economy of the process, an optimum enzyme concentration and an optimum cycle time can be obtained. An optimum use of the ultrafiltration plant will be achieved by operating two batch reactors in parallel with one separation unit. The repetitive batch mode has the advantage of being manageable in a simple way, so that it can be performed using standard equipment and so that the productivity can be maintained by simply supplementing, from time to time, with fresh enzyme. For the latter reason, the ultrafiltration unit has to be designed with a certain reserve with respect to flux.

If it is necessary to operate the membrane reactor continuously (UFMR) because of the scale of operation, the reaction is no longer uncoupled from separation. The principle design of a continuously operated membrane reactor has already been shown in Fig. 7. It is evident that the simplest UFMR will behave like a CSTR since a recycle stream has to be generated in order to avoid concentration polarization. Therefore, this simple type of UFMR should be used when the kinetics are characterized by a small K_m value and negligible product inhibition and/or when a relatively small degree of substrate conversion is required. In principle, a single UFMR can be operated at constant flux by feeding fresh enzyme according to the rate of deactivation. The latter method would lead to a flux decrease if the reactor is operated at constant pressure. When the productivity is given, it has to be achieved by maintaining a certain volumetric flow rate of substrate solution and maintaining a distinct conversion. It shall be assumed that the substrate concentration is fixed beforehand, so that its influence on the over-all process can be presumed constant. Since the feed rate is fixed, the only variable to maintain the desired conversion is the mass of enzyme in the reactor, according to Eq. (46).

The question now arises as to which reactor volume should be used to take up the required mass of enzyme. The reactor volume determines the enzyme concentration. On the basis of reaction engineering, the only answer is that the enzyme has to be maintained in a soluble state. The enzyme concentration does not interfere with productivity but with the layout of the required membrane area and the reactor surface. Since the pressure drop will increase when the enzyme concentration is increased at constant flux (see Fig. 2), a low enzyme concentration would be favoured. The lowering of the enzyme concentration is limited practically by the reactor cost because the reactor volume would have to be increased.

Up to now, only the effect of the concentration of active enzyme has been taken into account. But one of the main advantages of the UFMR is that fresh enzyme can be added easily to maintain the productivity in spite of catalyst deactivation. This leads to an increase in total protein concentration, though the amount of active catalyst remains constant. In consequence, the transmembrane pressure drop will rise with time until the maximum allowable pressure drop of the membrane is attained and the reactor has to be shut down for purging and reloading with catalyst. The time of operation, until shut-down is required, depends on the initial enzyme concentration and the deactivation constant. When the deactivation is assumed to follow a first order decay, the deactivation constant is defined as:

$$k_{de} = -\frac{1}{E}\frac{dE}{dt} \tag{52}$$

or to derive this for the supplementation by fresh enzyme as a function of the initial enzyme concentration:

$$-\frac{dE}{dt} = k_{de}E_o \tag{53}$$

In consequence the total concentration will increase linearly with time:

$$E_\Sigma = E_o(1 + k_{de}t) \tag{54}$$

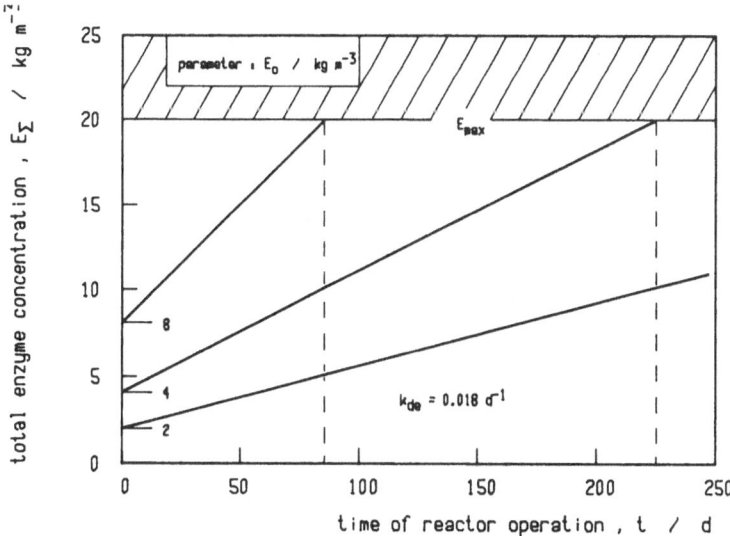

Fig. 19. The total enzyme concentration (active and deactivated) as a function of operating time for continuous membrane reactor operation

where

E_Σ kg m^{-3} total protein concentration (active + deactivated enzyme)

This is illustrated in Fig. 19. The maximum enzyme concentration is reached much more rapidly when the initial enzyme concentration is raised. The maximum operating time can be calculated:

$$t_{max} = \frac{E_{max} - E_o}{k_{de} E_o} \qquad (55)$$

where

t_{max} s   maximum operating time (until shut-down of the UFMR is needed)
E_{max} kg m^{-3} maximum total enzyme concentration (at which the maximum allowable pressure drop is reached at constant flux)

The maximum total enzyme concentration has to be evaluated experimentally. Therefore, the flux has to be measured at the maximum allowable pressure drop as a function of enzyme concentration. The appropriate hydrodynamic conditions have to be applied and maintained. Such an experiment will yield a relationship of the type shown in Fig. 20. This function must be modeled by linear or nonlinear regression for introduction in the following equations. For convenience it will be assumed that the interesting part of the plot can be described by the gel polarization model according to Eq. (10). The mass-transfer coefficient (k_d) is obtained from the slope of

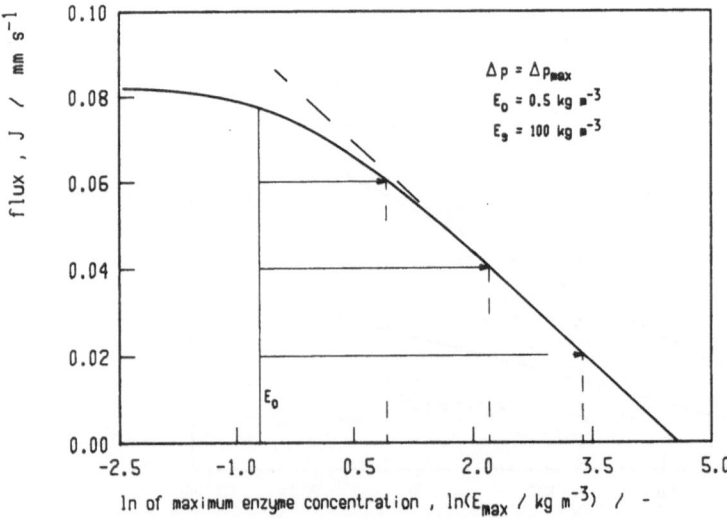

Fig. 20. Schematic representation of the achievable enzyme supplement for different transmembrane flow rates

the linear part of the curve in Fig. 20. Hence, the maximum total enzyme concentration can be derived as a function of flux:

$$E_{max} = E_s \, e^{-J/k_d} \qquad (56)$$

where

E_s kg m^{-3} enzyme saturation concentration

Replotting the information from Fig. 20 and idealizing according to the said assumption, yields Fig. 21.

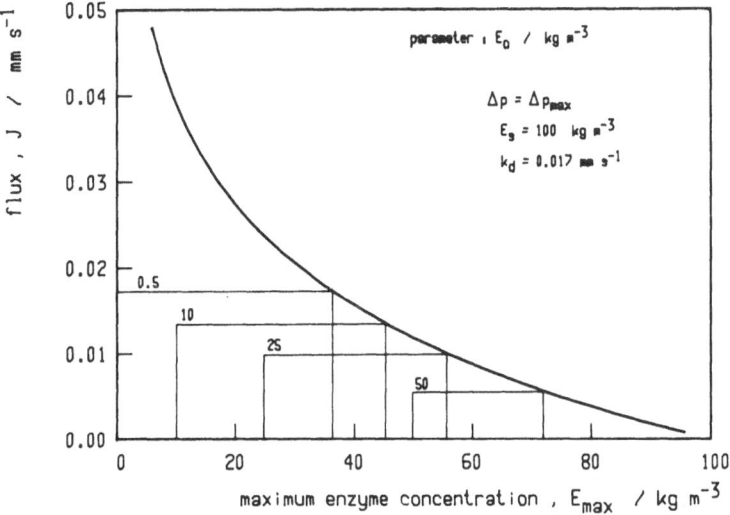

Fig. 21. The optimization strategy for membrane reactors for different initial enzyme concentrations

Here a difference on the enzyme concentration axis is directly proportional to the maximum operating time (t_{max}) for a given flux. It is evident that t_{max} will increase with decreasing flux and decreasing initial enzyme concentration.

The over-all objective will be to obtain a maximum amount of product during one operating period. This obviously means that the product of flux and maximum operating time should be optimized. With respect to Fig. 21, this means that a rectangle with a maximum area on the basis of the initial enzyme concentration (E_o) should be established. In conjunction with the gel polarization model, the objective function can be formulated by combining Eqs. (55) and (56):

$$(Jt_{max}) = \frac{V_f}{A_m} = J \, \frac{E_s \, e^{-J/k_d} - E_o}{k_{de}E_o} \qquad (57)$$

where

V_f m^3 total volume of permeate (filtrate) per operating period

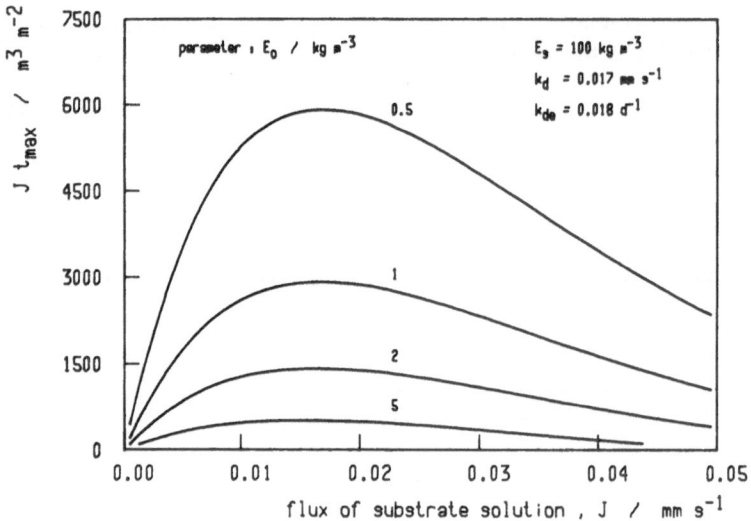

Fig. 22. The optimum productivity per operating cycle as a function of flux for different initial enzyme concentrations

This quantity is shown in Fig. 22, as a function of flux, for different initial enzyme concentrations. The over-all production per operating period exhibits an optimum value at a distinct flux. The optimum value decreases drastically with increasing initial enzyme concentration (E_o). With respect to Fig. 21, it is evident that the optimum flux must drop when E_o is raised. To calculate the optimum flux, Eq. (57) has to be differentiated with respect to flux and set to zero:

$$\frac{d(Jt_{max})}{dJ} = E_s \, e^{-J_{opt}/k_d}\left(1 - \frac{J_{opt}}{k_d}\right) - E_o = 0 \tag{58}$$

where

J_{opt} $m^3 \, m^{-2} \, s^{-1}$ optimum flux

This expression must be solved iteratively to obtain the optimum flux for a given initial enzyme concentration. The membrane area must be adjusted to achieve the desired productivity. A summary of the effect of E_o on the reactor performance is given for a fixed set of parameters in Table 3. The productivity is kept constant by increasing the membrane area, on the basis of the productivity, for an E_o of 1 kg m^{-3}. The loss of catalyst per unit time ($\Delta E/\Delta t$) due to shut-down of the reactor and discarding its catalyst content is also given. It turns out that the mass of enzyme needed to achieve the desired conversion should be diluted in an infinite reactor volume. Hence, the over-all optimization to yield a practical initial enzyme concentration depends on economic variables, particularly investment costs for reactor volume and membrane area, which have opposite dependence on E_o. Other criteria will also inhibit the unlimited decrease of E_o. It has often been found that enzymes are destabilized in dilute solution. To operate a reactor with large substrate/product content is a high economic risk, if contamination by microorganisms is feared and in the event of an uncontrolled shut-down of the reactor.

Table 3. Comparison of optimal membrane reactor operation as a function of the initial enzyme concentration

E_o kg m^{-3}	J_{opt} m^3 m^{-2} d^{-1}	E_{max} kg m^{-3}	t_{max} d	$(Jt)_{opt}$ m^3 m^{-2}	$\Delta E/\Delta t^*$ % d^{-1}	A_m m^2	V_R m^3	J_{opt}^{**} %
1	1.43	37.8	2043	2921	0.049	1.00	99.3	100.0
2	1.39	38.7	1020	1421	0.098	1.03	49.7	97.4
5	1.29	41.5	406	524	0.247	1.11	19.9	90.3
10	1.15	45.8	199	228	0.503	1.25	9.9	80.3
20	0.92	53.5	93	85	1.075	1.56	5.0	64.3
50	0.46	73	25.5	11.8	3.916	3.09	2.0	32.3

* loss of activity due to repeated shut-down of reactor $= 100/t_{max}$
** relative J_{opt}; relative productivity per unit membrane area
$E_s = 100$ kg m^3, $\quad \tau_m = 6000$ kg s m^{-3}
$k_d = 0.017$ mm s^{-1}, $\quad k_{de} = 0.018$ d^{-1}

The single stage UFMR will give, as discussed at the beginning of this chapter, the lowest utilization of enzymatic activity because it behaves like a CSTR.

To decrease backmixing but maintain the basic design, a stagewise operation of several UFMR in series can be envisaged. Substrate solution is fed to the first stage while a partially converted solution leaves it across the ultrafiltration module as permeate. The solution is pumped into the next stage and so on. For convenience, only cascades of equal-sized stages, each containing the same enzyme concentration, will be dealt with. The design with respect to flux is equal to that for a single UFMR. To calculate the productivity of such a cascade (CMR), several mass balances for single CSTR have to be solved in series. Therefore, Eq. (46) has to be extended to account for the case where the conversion is not zero in the feed downstream of the first stage:

$$\tau_m = \frac{m_{En}}{V_R} = \frac{1}{r_m} S_o(X_n - X_{n-1}) \tag{59}$$

where

τ_{mn}	kg s m^{-3}	space time for stage n
X_n	—	substrate conversion in stage n
X_{n-1}	—	substrate conversion in feed of stage n
m_{En}	kg m^{-3}	mass of enzyme in stage n

It has to be noted that the reaction rate r_m in stage n depends on the conversion X_n. When the feed rate $(\dot{V}_R)$ is given, the total mass of enzyme has to be calculated. This can be done by solving Eq. (59) for each stage and varying m_{En} until the desired conversion is obtained after the last stage of the cascade. The productivity per unit catalyst is then with respect to the assumption made:

$$\bar{r}_m = \dot{V}_R \frac{S_o X_n}{n m_{En}} \tag{60}$$

where

n — number of stages

A three stage CMR yields a much better productivity compared to a single UFMR, as shown in Fig. 23, for the kinetics discussed in conjunction with Fig. 17. However, this concept suffers from a severe disadvantage. Since each stage has to be equipped with a feed pump, a complete ultrafiltration unit and the facility for dosing additional catalyst, such reactor combinations will require considerably higher investments than does a single stage UFMR.

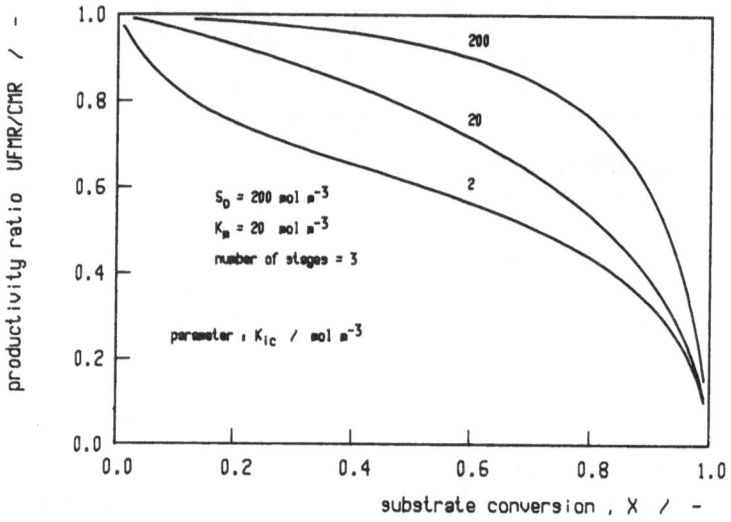

Fig. 23. Productivity comparison of a single UFMR with a cascade of membrane reactors (CMR) as a function of substrate conversion

This problem can be overcome by designing the cascade in the way shown in Fig. 24. A simple cascade of stirred tanks is used in series with only one separation unit. The catalyst is retained in the separation unit, concentrated and recycled to the first stage. The advantage of this design is that standard reactors can be used in conjunction with only one separation unit. The calculation of the over-all productivity is somewhat more tedious than for normal cascades. The enzyme concentration in

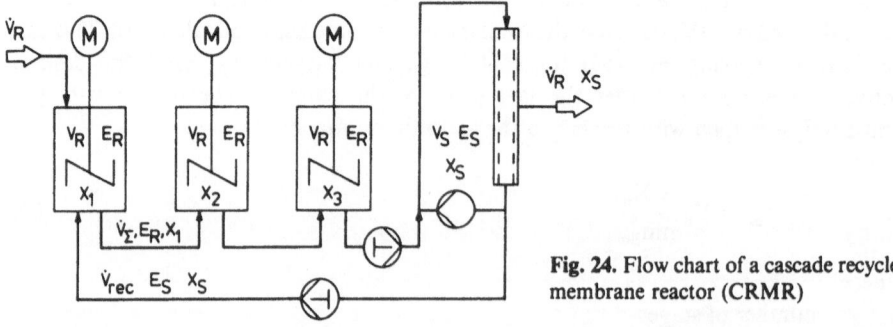

Fig. 24. Flow chart of a cascade recycle membrane reactor (CRMR)

the n stirred tanks will normally differ from that in the separation unit. Both depend on the recycle ratio (see Fig. 24):

$$f_{rec} = \frac{\dot{V}_{rec}}{\dot{V}_R}$$
(61)

and the volume ratio:

$$f_V = \frac{V_S}{V_\Sigma}$$
(62)

where

f_{rec}	—	recycle ratio
$\dot{V}_{rec}$	$m^3 \ s^{-1}$	volumetric recycle flow rate
$\dot{V}_R$	$m^3 \ s^{-1}$	volumetric feed rate
f_V	—	volume ratio
V_S	m^3	volume of separation unit
V_Σ	m^3	volume of whole reactor system

The main balances for the enzyme are:

$$m_{\Sigma E} = nV_R E_R + V_S E_S = V_\Sigma E_\Sigma$$
(63)

and the steady state flow:

$$\dot{V}_{rec} E_S = (\dot{V}_R + \dot{V}_{rec}) E_R = \dot{V}_\Sigma E_R$$
(64)

where

$m_{\Sigma E}$	$kg \ m^{-3}$	total mass of enzyme
n	—	number of stirred tanks
V_R	m^3	volume of a stirred tank
E_R	$kg \ m^{-3}$	enzyme concentration in the stirred tanks
E_S	$kg \ m^{-3}$	enzyme concentration in the separation unit
E_Σ	$kg \ m^{-3}$	global enzyme concentration
$\dot{V}_\Sigma$	$m^3 \ s^{-1}$	volumetric flow rate passing the installation ($= \dot{V}_R + \dot{V}_{rec}$)

Out of these expressions, the enzyme concentrations can be calculated to:

$$E_R = \frac{f_{rec}}{f_{rec} + f_V} E_\Sigma$$
(65)

$$E_S = \frac{1 + f_{rec}}{f_{rec} + f_V} E_\Sigma$$
(66)

For the determination of the individual catalyst mass referred space time, the mean residence times of the reactants in each part of the reactor have to be known:

$$\bar{t}_R = \frac{1 - f_V}{n} \bar{t} \tag{67}$$

$$\bar{t}_S = f_V \bar{t} \tag{68}$$

where

$\bar{t}_R$ s   mean residence time in a stirred tank
$\bar{t}_S$ s   mean residence time in the separation unit
$\bar{t}$ s   over-all mean residence time ($= V_\Sigma/V_\Sigma$, referred to a single passage of the reactor)

These expressions lead to the space times:

$$\tau_{mR} = \frac{1 - f_V}{n} \frac{f_{rec}}{f_{rec} + f_V} \tau_m \tag{69}$$

$$\tau_{mS} = \frac{f_V(1 + f_{rec})}{f_{rec} + f_V} \tau_m \tag{70}$$

where

τ_{mR} kg s m^{-3} mass referred space time in each stirred tank
τ_{mS} kg s m^{-3} mass referred space time in the separation unit
τ_m kg s m^{-3} mass referred space time ($= E_\Sigma \bar{t}$, referred to a single passage of the reactor)

The over-all mass balance is solved by calculating the conversion according to Eq. (57) stage by stage, introducing the space times as defined in Eqs. (69) and (70). Since the recycle stream contains not only enzyme but also product, recycling leads to a partially converted feed:

$$X_o = \frac{f_{rec}}{1 + f_{rec}} X_S \tag{71}$$

where

X_o — substrate conversion in feed due to recycling
X_S — conversion in the separation unit (final conversion)

This has to be taken into account for a repeated calculation of the whole reactor. It is repeated until the final conversion no longer changes.

The productivity per unit mass of enzyme is evaluated taking into account the over-all space time of substrate:

$$\tau_{m\Sigma} = E_\Sigma \frac{V_\Sigma}{V_R} = \tau_m(1 + f_{rec}) \tag{72}$$

where

$\tau_{m\Sigma}$ kg s m^{-3} catalyst mass referred space time (productivity referred)
to yield

$$\bar{r}_m = \frac{S_o X_s}{\tau_{m\Sigma}} \qquad\qquad (73)$$

For the cascade reactor with catalyst recycling (CRMR) the improvement in performance compared to an equal-sized single UFMR containing the same amount of enzyme
is primarily determined by the volume ratio (f_V) and the recycle ratio (f_{rec}). The
ratio of separation unit volume to total volume (f_V) has to be achieved. This implies
enzyme can be used most efficiently in the cascade of stirred tanks. Therefore, a low
a high ratio of membrane area to separation unit volume. The influence of the
recycle ratio is more complex. If there is no recycle stream, the total mass of enzyme is
concentrated in the separation unit yielding a single low-volume CSTR. When the
recycle flow rate is more than 20 times higher than the feed rate, all concentrations
will be approximately equal, thus yielding a single CSTR of high volume. Since only
the mass of enzyme per unit feed rate, the catalyst mass referred space time
(τ_m), determines the reaction rate, these two cases will yield the same productivity
as any other single CSTR. The cascade with recycling will work optimally when
backmixing is minimized ($f_{rec} \to 0$) and the enzyme is found predominantly in the
stirred tanks ($f_{rec} \to \infty$). Thus, at intermediate recycle ratios an optimum productivity
is obtained as illustrated in Fig. 25. By simple optimization strategies, the optimum
recycle ratio with regard to the instantaneous optimum productivity can easily be
calculated.

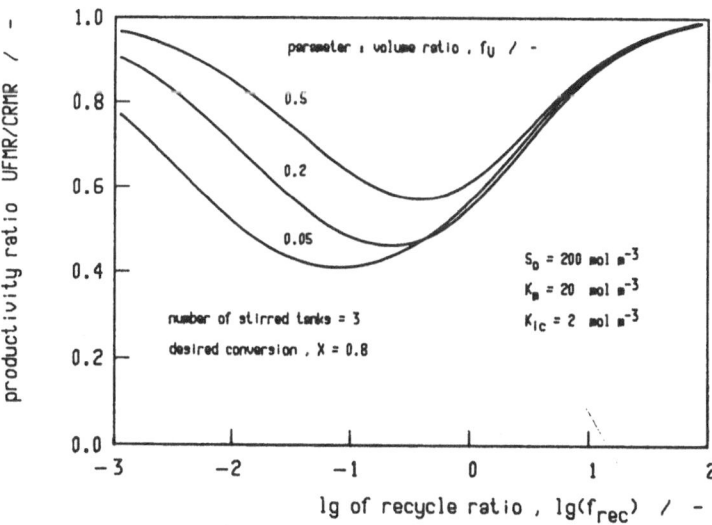

Fig. 25. The influence of the recycle ratio of a CRMR on the performance ratio compared to a single
UFMR

Since all membrane reactors have to be shut down after a certain time of operation due to the accumulation of deactivated catalyst, the intrinsic recycle ratio must be evaluated taking into account the optimum productivity per operating period, as defined by Eq. (58). Since all types of reactors with catalyst recycling contain a higher enzyme concentration in the separation device compared to single UFMR of equal volume, the flux and the operating period will be reduced according to Fig. 21. In principle, the improvement with respect to backmixing must be paid for by a larger membrane area or a shorter operating period at comparable enzyme concentrations. An entire economic analysis would show that the CRMR is superior to single UFMR, but that it has to be operated at lower over-all catalyst concentrations.

Stagewise processing is always used to approximate plug flow behaviour. Since in practice, the number of stages is limited to 3 or at maximum 5, due to economic constraints, a tubular design would be favoured. On the one hand backmixing could probably be suppressed to an extent corresponding to more than 5 stages, thus improving the utilisation of catalytic activity. On the other hand, movable parts, necessary in cascades of stirred tanks, would be completely avoided, thus improving economics by decreasing investment costs, facilitating maintenance and reducing power consumption.

The flow chart of such a tubular recycle membrane reactor (TRMR) is given schematically in Fig. 26. It operates similarly to the CRMR but with the difference that the cascade is replaced by a tubular reactor. The main design parameter besides those already discussed in connection with the CRMR is to decrease backmixing in the tubular reactor. It is known that tubular reactors with large diameters behave far from ideal plug flow behaviour especially at low superficial fluid velocities. The reason for this results mainly from superimposed circulation of fluid by free convection. To avoid this flow instability and smooth the radial concentration profile, special motionless mixers (e.g. SULZER motionless mixers type SMV) can be introduced into the tube. Thus, backmixing can be diminished and will correspond to 5 to 40 equivalent stirred tanks per meter of reactor length depending on the tube diameter. The TRMR performance may be calculated either exactly as has been demonstrated in case of the CRMR taking into account the number of equivalent stages over the total tube length or, if the number of stages exceeds approximately

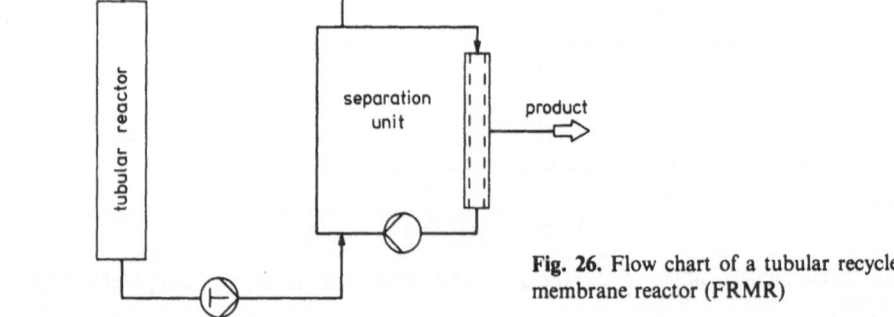

Fig. 26. Flow chart of a tubular recycle membrane reactor (FRMR)

25, by integration of the over-all mass balance equation of the tubular reactor according to Eq. (49). The design with respect to flux and the optimization of the operation are essentially the same as has already been discussed for the CRMR.

For the discussion of membrane reactors, it has been assumed thus far that the catalyst will be the only component of the reaction system which is retained by the ultrafiltration membrane. If there are high molar mass or particulate impurities, especially in the substrate solution, these would lead to a drastic diminution of the operating period. Therefore, these troublesome impurities must be prevented from reaching the separation device. This can be accomplished by forcing the substrate solution through sterilizing filters or — better and if necessary — through an ultrafiltration unit prior to the reactor. On the other hand, membrane reactors have found particular interest to convert macromolecular substrates (e.g. proteins, starch, cellulose) to their monomers or oligomers owing to the absence of diffusional restrictions. Diffusional limitations do not allow these substrates to be effectively converted by insolubilized enzymes. Operating membrane reactors with such substrates means that the substrate is retained. Furthermore, such substrates may contain impurities (e.g. retrograded starch) which cannot be widthdrawn prior to the reaction. The retention of substrate requires an extension of the reactor mass balance equations and it has to be assured during continuous operation that a steady state is attained at a low substrate level, otherwise, the flux would be drastically reduced. The greatest drawback are the impurities which tend to diminish the operating period.

When this problem is very severe and the enzyme cannot be withdrawn simply after the reaction, the concept of the porous wall tubular reactor (PWTR) might lead to a reasonable compromise. This reactor has been proposed and the performance theoretically predicted by Shah and Remmen [62] and later applied to enzymatic processing (see Chapter 5). The principle of this concept is to use a tubular ultrafiltration device which is fed at one end with substrate and enzyme solution. The substrate and the enzyme are retained whilst the products pass through the membrane. The product solution is forced under pressure through the membrane wall whilst the enzyme, unconverted substrate and any troublesome impurities leave the system at the other end of the tubular ultrafiltration device. This technique keeps the catalyst in contact with the substrate for a longer time than it would be the case for a solid wall tubular reactor.

The PWTR as described is unlikely to find industrial application since it suffers from the drawback that concentration polarization is extremely severe due to low superficial velocities. The flow rate is coupled to the residence time of the reactants and it even decreases with reactor length due to the continuous removal of product solution. This problem could be overcome by uncoupling the superficial fluid velocity along the membrane from the mean residence time. The reactor could be designed as a cascade for continuous concentration as shown in Fig. 13. Here the hydrodynamic conditions can be chosen for each stage in order to attain a reasonable over-all flux of product solution.

If it is not necessary to operate the reactor continuously, the best solution would be to employ the repetitive batch operating mode. The catalyst could be recycled several times until the time required for ultrafiltration would exceed reasonable limits. Then the catalyst would have to be discarded, the membrane cleaned and the reactor recharged with fresh catalyst.

4 Application of Ultrafiltration for the Isolation of Biocatalysts

For biocatalyst isolation there are three main areas, where ultrafiltration could be employed advantageously: concentration, diafiltration (substitute for dialysis) and fractionation according to size. Whilst for reasons discussed below, ultrafiltration has not yet met expectations with respect to fractionation, the first two operations are presently conducted by ultrafiltration at both laboratory- and production-scale. Owing to the particular characteristics already mentioned in Sect. 1.1, ultrafiltration compares favourably with competing classical processes such as precipitation, lyophilization etc. The improved chemical resistance of membranes and ancillary parts has made periodic cleaning and sterilization possible, thereby considerably improving the lifetime of ultrafiltration membranes.

Depending on the particular circumstances, average service times for membranes of up to several years are reported [63]. There is some scepticism about such claims, but a maximum service life of one year is expected [64].

In the following chapters, general features of ultrafiltration processes for enzyme recovery and purification will be discussed together with examples from recent literature.

4.1 Process Variables of Main Influence

As discussed in Chapter 2, general mathematical treatment of the ultrafiltration process based on mass transfer considerations is well developed and has led to improvements in the design of equipment to control concentration polarization. Less well understood are influences arising from solution properties such as pH, salt concentration and presence of uncharged small molecules. These parameters can be changed within certain limits in enzyme processing, if the biological activity and stability are not impaired. Two mechanisms have to be considered to account for such influences: interaction with the polymer matrix forming the active layer in the commonly employed anisotropic membranes and interactions with the macromolecular solutes.

Interactions of salts with the polymer matrix was demonstrated by Wang et al. [65] resulting in a marked decrease in the initial filtration rate. This decrease depends on the kind and concentration of ions used and also on the charge of the anion. These results suggest that low ionic strengths are favourable and that multivalent anions should be avoided to save processing time. Indeed it is often found beneficial to dilute a batch prior to concentration. It may also be anticipated that solutions containing nucleic acids, as encountered when handling intracellular enzymes, are more difficult to process. Some partitioning of small ions has been observed at ultrafiltration membranes [65, 66, 67] but not of uncharged small solutes like ethanol and sucrose [66, 67, 68]. This has to be taken into account if removal of salt from process liquors is calculated according to the permeate concentration.

Adsorption of proteins or other constituents to the polymer matrix vary for different membranes and have to be considered for process development. One manufacturer quotes adsorption of 1.5–300 $\mu g\,cm^{-2}$ of albumin by contact to membranes of different polymer composition. Approximately the same order of magnitude can be expected for membranes from other sources. Adsorption is non-specific and influenced by solution parameters, hydrophobic molecules are preferentially bound. Even for protein bulk concentrations as low as 1 µg ml^{-1} adsorption resulted in a drop in permeate flow by 37%[69]. Due to concentration polarization, the actual adsorbate concentration on the surface of the membrane by far exceeds the bulk concentration of dilute solutions. Adsorption may also affect the recovery of products. This point has to be considered in cases where minute amounts of biologically active proteins have to be concentrated. In such circumstances, the membrane area should be reduced to a practical minimum size.

Interactions between environmental factors, like pH and ions, with proteins and their possible consequences on ultrafiltration have been studied by Melling[45]. As shown in Tab. 4, ultrafiltration characteristics of enzymes can vary significantly with pH. There is also a very pronounced and differential effect of various ions, resulting in a markedly changed retention of *Staphylococcus aureus* penicillinase using ultrafiltration membranes with a different cut-off[45]. These effects might be due to changing charge, hydration and solute-solute interaction as well as changes, however small, in molecular shape and size of the proteins.

Table 4. The effect of pH on the concentration process of enzymes out of culture broths (Melling, J.: Process Biochem., p. 7, Sept. 1974, with kind permission of Wheatland Journals Ltd.)

Parameter	pH	Integral retention %		Ratio of specific activity	
Membrane		PM-30	XM-100	PM-30	XM-100
E. coli	5.1	58	34	5.9	3.1
penicillinase	5.6	54	18	7.2	3.6
	6.1	50	17	7.4	3.0
	6.7	66	18	4.4	2.1
	7.1	72	19	4.4	2.1
	7.8	80	25	3.2	1.7
S. aureus	4.0	23		1.2	
penicillinase	5.0	57		0.8	
	6.0	71		0.6	
	6.5	89		0.98	
	7.0	94		1.0	
	8.0	100		1.1	

Membranes: PM-30 cut-off 30,000 g mol^{-1} (Amicon)
XM-100 cut-off 100,000 g mol^{-1} (Amicon)
Enzymes: *E. coli* penicillinase, M = 22,500 g mol^{-1}
S. aureus penicillinase, M = 20,600 g mol^{-1}

Therefore, electrolyte concentrations and pH are important parameters which should be controlled in order to achieve reproducible and optimal performance of ultrafiltration plants. If certain ions like Ca^{++} are known to stabilize the tertiary structure of a protein, these ions have to be supplied during diafiltration to avoid excessive activity loss. Such effects have been reported for example by Heinen and Lauwers[70] during ultrafiltration of amylase from *Bacillus caldolyticus*. It should also be mentioned that the choice of pH and the concentration of ions as well as water miscible solvents like alcohol will influence the solubility of proteins. This is important with respect to concentration polarization since its effect is intensified by lowering the saturation concentration of solutes. Near the isoelectric point, where the solubility of proteins is minimal, the flowrate drops significantly. In complex mixtures, the component with the lowest solubility will have a dominating influence on flux. Therefore, it is advisable to select conditions which maximize protein solubility. If necessary, precipitating agents should be removed by diafiltration prior to concentration in order to save processing time.

Temperature is an important independent variable in ultrafiltration processes. It is limited on one hand by the thermal stability of the product, and on the other hand by the requirement to control microbial contamination. Reduction of the operating temperature can sometimes be avoided if residence times are sufficiently short in continuous processes. Otherwise a decrease of 1–3% in capacity has to be accepted for every °C [71]. Several parameters determining the flux through ultrafiltration membranes also depend on temperature. Investigations by Pace et al. [68] on the influence of viscosity and temperature on flux, have demonstrated that changes in flux with temperature could be largely accounted for by changes in viscosity. Alteration in flux due to changes in diffusivity of nonpermeating species with temperature are found to be comparatively small, thus, explaining the observed correlation of flux with the bulk viscosity of the process solution. Concentration of extracellular enzymes on an industrial scale is reported to be carried out at 10 °C which seems to be a compromise between the achievable flux and enzyme stability [63].

4.2 Concentration of Biocatalysts

For the isolation and purification of enzymes, several concentration steps are usually necessary. These are primarily designed to reduce the volume of the solution to be handled, transported or stored. Especially on the industrial scale, ultrafiltration seems to have largely replaced the salt- or solvent-precipitation steps previously employed. Ultrafiltration now precedes energy intensive processes like spray drying or lyophilization. The highest absolute protein concentrations achievable by ultrafiltration depend mainly on the solubility of the solutes. For blood fractionation, the final albumin concentration obtained in thin channel and cassette systems is reported to be 40%, in hollow fiber cartridges concentrations up to 25% albumin can be achieved, but processes tend to be slow [72]. For the concentration of extracellular enzymes from culture broth, final solid contents of the process solutions are reported to be within the range of 11–19% [63]. Since culture broths contain not only extracellular enzymes but other unidentified material, the situation is much more complex in comparison to the rather clean system of albumin. Activity yields of enzymes are usually reported to be around 90%. Provided a correct membrane is selected, the final yield will depend not only on the intrinsic stability of the enzyme towards the solution parameters but also on their sensitivity to denaturation at interfaces, and to shear. The latter effect has been discussed in Sect. 2.3.

Besides the desired concentration of the product, the simultaneous removal of contaminants of smaller molar mass is of considerable benefit for further processing [63]. The removal of sugars, amino acids and peptides, which are the major cause of discoloration, hygroscopicity, odour, gumming, and caking, is essential to avoid deterioration of products. Table 5 lists some results and serves to illustrate the scale of industrial operation [63].

Table 5. Concentration of enzymes out of culture broths
(Neubeck, C. E.: U.S. 4,233,405, Nov. 11, 1980)

Enzyme	Source	Initial mass t	Final mass t	Final solids %	Final volume %	Yield %
α-Amylase	B. subtilis	14	3.2	33	22.9	82
α-Amylase	B. subtilis	18.6	3.9	25	21.0	88
Neutral protease	A. flavus	1.6	0.2	15	12.5	90
Acid protease	A. oryzae	59	4.4	13	7.5	90
Neutral protease	B. subtilis	~2.6	0.4	16	15.4	95

If possible, a membrane is chosen with the highest cut-off sufficient to keep the enzyme of interest in the concentrate. This will allow not only the removal of solvent and low molar mass solutes during concentration, but may achieve considerable purification. For example, extracellular acid phosphatase from *Saccharomyces cerevisiae* has been concentrated using XM-300 membranes (cut-off $\sim 300,000 \text{ g mol}^{-1}$) resulting in a 26fold increase in specific activity. 97 % of the bulk protein was removed by this step [73]. Hummel et al. [74] employed a hollow fiber cartridge with an exclusion limit of $80,000 \text{ g mol}^{-1}$ for the large scale isolation of leucine dehydrogenase from *Bacillus sphaericus* removing 50 % of contaminating proteins with the permeate.

During any batch volume reduction, the concentration of the non-permeating species will increase leading to a reduction in flux. Howell et al. [69] discuss the decrease in flux at various time scales during ultrafiltration. Concentration polarization occurs very fast, <5 s, followed by a further rapid drop in flux within 10 min due to build up of the gel concentration at the membrane. A third phase is observed where the gel layer resistance increases by a slower rate for several hours. A second order relationship was found for the gel layer growth rate with respect to the protein concentration at the wall. In complex mixtures, the change of flux with time is less predictable and greatly influenced by the properties of escort substances. Therefore, trial experiments are unavoidable to obtain reliable data for plant design. An exponential decline of flux as a function of time has been reported for batchwise enzyme concentration [75, 76] as well as a linear decline [77]. This dependence cannot be generalized since it is influenced by the initial conditions. When the initial volume is small, the flux decline will, for the same initial flux, be steeper than in the case of higher initial volumes. The concentration of a fungal protease in the culture broth has been studied [78] using a Millipore PTG6 10,000 cassette ultrafiltration system of 0.92 m². The initial volume of 300 l has been reduced to 10 l during 20 h of batch operation (recycle mode). The operational conditions were: a constant pressure of 3.5 bar and a recycle flow rate of 90 l h^{-1}. 96 % of the activity has been recovered while only 39 % of the proteins were retained. The actual enzyme concentration during ultrafiltration is a linear function of the volume concentration factor V_0/V according to Eq. (30).

The flux, as a function of the natural logarithm of V_o/V, is given in Fig. 27 to test the applicability of the gel polarization model. The linearization of this model yields an initial flux $J_o = 21.22 \, l \, h^{-1} \, m^{-2}$ and a mass-transfer coefficient of $k_d = 4.96 \, l \, h^{-1} \, m^{-2}$. Inserting these values in Eq. (33), a process time of 20.4 h is calculated which is in close agreement with the experiment. To demonstrate the flux decline during ultrafiltration, the flux is shown in Fig. 28 as a function of operating time. Owing to the high initial ratio of volume to flux, the volume concentration factor increases only slightly. This leads to a small decline of flux which is intensified when the total volume

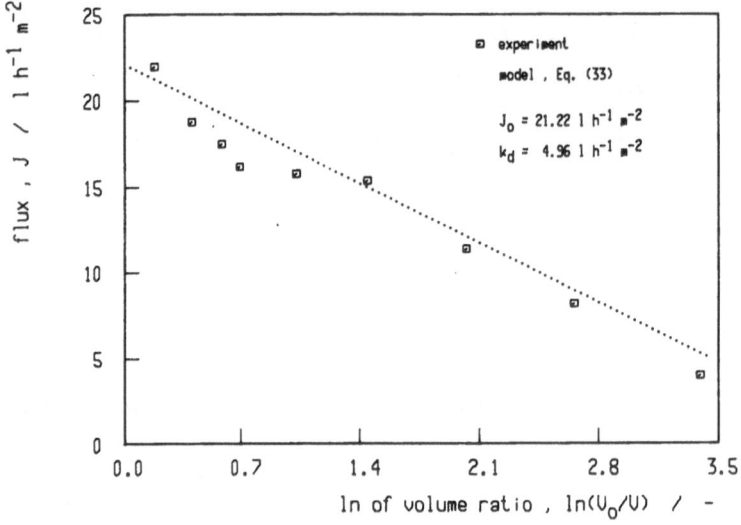

Fig. 27. Test for the validity of the gel polarization model [78] (for experimental details see text)

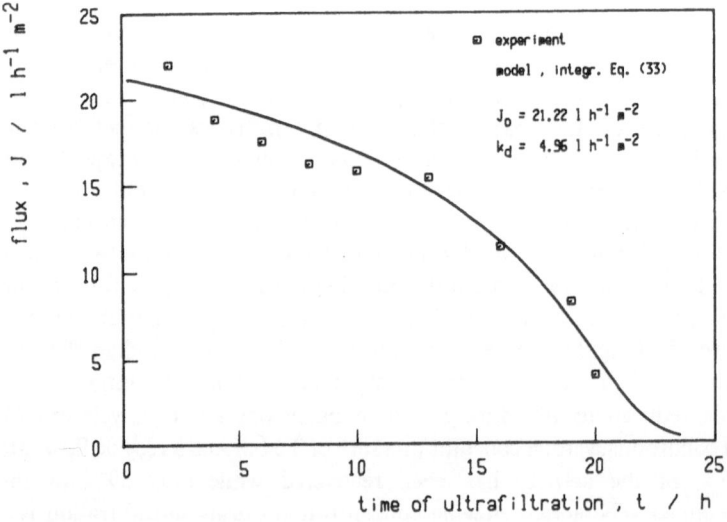

Fig. 28. Flux as a function of operating time [78] (for experimental details see text)

diminishes. The values obtained for the initial flux (J_o) and the mass-transfer coefficient (k_d) have been utilized to scale up the process for concentrating 3000 l of culture broth by means of a 4.6 m² module.

While ultrafiltration is extensively employed for enzyme processing, the recovery of biomass still seems to be in the phase of development.

Batch and continuous operation of diverse laboratory-scale ultrafiltration devices have been studied for the concentration of *Escherichia coli* [79]. Since the aim has been directed to bacteriological analysis, recovery is stressed as the primary objective. Near to 100 % recovery has only been achieved in batch operation employing capillary modules and by backflushing the membrane in the reverse flow sense. In addition a prefilter had to be used.

The production of bakers yeast on deproteinized whey in a two-stage process has been described [80]. The *Streptococcus thermophilus* biomass was removed after the lactic fermentation by ultrafiltration through a capillary module (GM 80, ROMICON) before the permeate could be re-used for the process by *Saccharomyces cerevisiae*. Unfortunately no details on biomass recovery are reported.

Ultrafiltration will find application where high product purities are required such as in vinegar production. A so-called cross-flow filtration has been reported [24, 25] for the separation of bacteria in a vinegar process. It works at relatively low pressure under laminar flow conditions. A brief theory is given to account for the required flow rate along the membrane to keep the microorganisms in steady movement so as to avoid plugging of the membrane pores. As in conventional filtration, the addition of filter aids favour the achievable flux. In cross-flow filtration, they serve to entrain the bacteria so that these do not adhere to the membrane.

4.3 Diafiltration and Fractionation

During protein purification, defined salt exchange may become necessary prior to chromatographic separations. Also the complete removal of solutes with low molar mass is often desirable either during the final stages of purification or prior to lyophilization. For such purposes, diafiltration is currently replacing the time honoured process of dialysis, if large volumes have to be handled. Due to enforced flow across the membrane, the solute exchange by diafiltration is speeded-up by a factor of about 30 compared to conventional dialysis. In addition, diafiltration can be conducted in such a way as to minimize the consumption of the equilibration solvent. Two process modes seem to be preferred (see Sect. 3.1): diafiltration by a series of concentration and dilution steps and diafiltration at constant volume.

In the concentration-dilution mode, the concentration of small solutes is constant at each stage while the protein concentration increases during each concentration step.

Working with constant volume, the bulk protein concentration will not essentially change. The washout of small solutes is then directly proportional to the flow rate. The consumption of eluant is a function of the initial volume and does not depend on the protein concentration. Prior concentration may reduce the volume of buffer required and also shorten the diafiltration time which, on the other hand, depends on protein concentration. Ng et al. [67] demonstrated that an optimum bulk protein concentration exists, 7–8 % in the case of albumin, which leads to a minimum processing time. Martinache and Henon [72] optimized three parameters: bulk concentration of proteins, consumption of buffer, and processing time for the removal of ethanol from albumin solutions. The following procedure was adopted for processing 1,000 l of a 2% albumin solution containing 40% ethanol: dilution to 20% ethanol concentration, thereby increasing flux; concentration to about 9% protein results in volume reduction and ethanol removal; diafiltration at constant volume using 2000 l water, thereby lowering the ethanol concentration to 0.03%; and a final concentration to 20% protein.

In this special case, processing time is cut and buffer consumption is only half that in a concentration-dilution process. Similar strategies can be followed to optimize time and buffer consumption for enzyme processing, but such studies have not yet been described in detail in the literature.

In principle, ultrafiltration membranes are expected to allow protein separations according to size. This theoretical potential is, to a large extent, abolished by concentration polarization. Accumulating retained solutes at the membrane surface alters the retention characteristics of the membrane as well as the flux. While flux can be improved by effective measures that control the polarization, the fractionation potential is still impaired by the remaining polarization layer which cannot be completely avoided. The polarization layer acts as an additional filter with properties which are essentially unpredictable for real systems. The situation is further complicated by interactions between retained and eluting solutes.

The passage of the eluting component will be decreased and fractionation hindered. Model studies for staged ultrafiltration processes have shown that separation and purification can be improved in this way [43, 44], but have to be carried out at extremely low protein concentrations which are unrealistic for practical purposes.

The complete separation of two or more components in high yield by ultrafiltration remains a problem that is not likely to be solved in the near future for cases where the molar masses do not differ by a factor of 10 or more. However, some degree of modest fractionation can be achieved by ultrafiltration by choosing appropriate variables during the concentration or the diafiltration of enzymes in complex protein mixtures.

If the potential of ultrafiltration is combined with other means, effective fractionation can be achieved. Favourable results can be obtained by utilizing the interaction of enzymes and other macromolecules. Thus, it has been found that pullulanase can be purified in the presence of DEAE-dextran [78]. Pullulanase alone is retained by Amicon XM-300 membranes only to 13%. In the presence of a buffer of low ionic strength that contains 1% DEAE-dextran, the retention increases to more than 96%. The enzyme can be dissociated from the DEAE-dextran by increasing the ionic strength of the phosphate buffer above 0.1 M and, in consequence, is no longer retained by the membrane.

Another example of a specific interaction has been described by Bartling and Barker [81]. They fractionated horse-radish peroxidase ($M = 42,000$ g mol^{-1}) and bovine pancreatic trypsin ($M = 32,200$ g mol^{-1}) by means of ultrafiltration applying crosslinked soybean trypsin inhibitor of high molar mass. At pH 8.3, the peroxidase could be eluted through a membrane with a cut-off of 50,000 g mol^{-1}. When the pH was shifted to 1.8, trypsin was eluted, whilst the crosslinked inhibitor was retained for further use.

5 Application of Ultrafiltration for Membrane Reactors

A literature survey is given concerning the application of pressure driven processes for biocatalytic reactions. The examples are summarized according to the biocatalysts used and the main findings are discussed. The subheadings are chosen according to a classification in groups of increasing reaction complexity.

A vast number of articles are available dealing with immobilization of enzymes by an ultrafiltration membrane, utilizing a concentration gradient as the driving force for mass transfer (e.g. dialysis reactor). Those who are interested in these techniques can refer to the review of Chambers, Cohen and Baricos [82] or for a more general classification, to Flaschel and Wandrey [34].

5.1 Enzyme Systems

The first application of ultrafiltration for the physical immobilization of enzymes was reported by Blatt et al. [10] in 1968. This technique has been evaluated more recently for quite a number of different enzymes, especially hydrolases. To facilitate the discussion, this subject is subdivided according to the molar mass of the substrates employed.

The applications of single enzyme systems in ultrafiltration membrane reactors to convert low molar mass substrates are summarized in Table 6. Normally the enzymes are employed in the native state without prior modification. If "modified" is stated, this means that the enzyme has been artificially enlarged by crosslinking [86, 105] adsorption and binding to soluble polymers [86, 96, 102] or by adsorption to bentonite [95]. One of the reasons for applying these techniques is to enlarge the particle size of small enzymes. Thus, high flux membranes which are not sufficient to retain the native enzyme efficiently can be employed. On the other hand, it serves to stabilize enzymes generally [108, 109] but especially in the case of proteolytic enzymes [86, 102, 105] against autolysis.

The preferred reactor design employed is the commercially available stirred ultrafiltration cell equipped with a single flat membrane. This reflects that most of the applications cited have been studied on laboratory-scale. Two examples have been studied on pilot-scale employing a capillary ultrafiltration module of 2.8 m^2 [88] and a plate-and-frame module of 18 m^2 [91]. Since the reactors are preferentially operated in the continuous mode applying either constant pressure or constant flux, there are only a few examples of behaviour different from continuous stirred tank reactors (CSTR). The operation of a 2-stage cascade has been reported [86] while a 3-stage cascade of membrane reactors has been simulated [84]. In the latter example, product solution is collected downstream of the reactor and is applied once again to the same reactor. Although thin-channel devices have been employed in series with a stirred tank [95, 107], their cascade behaviour is negligible since the ultrafiltration device is too small to have a significant effect on backmixing. The only study of a cascade with catalyst recycling is given for the optical resolution of methionine by means of acylase [83]. A 2-stage cascade of stirred tanks is connected in series to a separation device. The achieved improvement, as compared to a single stage, corresponds well to the prediction of the model.

Tables 6. Enzymes employed in membrane reactors to convert low molar mass substrates

Enzyme	Source	Native	Modified	Substrate	Repet. batch	Continuous	Cascade	Staged with cat. recycling	Δp const.	J const.	Flat membrane	Tubular membrane	Lab.-scale	Pilot-scale	Ref.
Acylase	Hog kidney	×		N-Acyl amino-acids		×		×		×		×	×		46, 38)
Cellobiase	T. viride	×		Cellobiose		×	s				×		×		84)
Cephalosporin acetyl esterase	B. subtilis	×		7-ACA	×				×		×		×		85)
α-Chymotrypsin	Bov. pancreas	×	×	Phe-methyl ester		×	×			×	×		×		86)
Formate dehydrogenase	C. boidinii	×	×	Formate/NAD		×				×	×		×		87)
β-Galactosidase	K. lactis	×		Whey permeate		×						×		×	88–90)
β-Galactosidase	K. fragilis	×		Whey permeate	×					?	P			×	71, 91)
Invertase	Yeast	×		Saccharose		×			×		×		×		92)
Invertase	Yeast	×		Saccharose		×				×	T		×		93, 94)
Penicillin amido hydrolase	B. megaterium	×	×	Benzyl penicill.		×			×		×		×		95)
Penicillin amido hydrolase	—		×	Benzyl penicill.	×		*				×	×	×		96)
Xylose isomerase	P. hydrophile	×	×	Glucose	×					×	×		×		166)

s   simulated
*   claimed, no example
~   supposed
P   plate-and-frame module, T thin-channel device

The repetitive batch mode is preferentially applied for reactions which need pH-control [85, 96] or a long reaction time due to low temperatures and product inhibition [91, 166]. On the other hand, it is no problem to control the pH in continuously operated membrane reactors since their content is homogeneously distributed due to vigorous agitation. In many cases, it might be favourable to adjust the pH of the feed solution just before entering the reactor in such a way that the optimum pH for the reaction system is attained in the reactor at steady state. This strategy has been demonstrated by Weiss [86] for a single UFMR as well as for a 2-stage cascade with pH-adjustment before each stage.

The applications cited will be discussed in terms of catalyst retention, operational stability and the effectiveness of the enzymes. During multiple re-use studies, Abbott et al. [85] stated 50% loss of protein and activity after 20 repetitive batch cycles. Since the molar mass of the esterase used is reported to be $\sim$190,000 g mol^{-1}, it seems to be very unlikely that it passes a PM 10 (cut-off 10,000 g mol^{-1}) membrane. Either the esterase is degraded or cleaved into subunits or this loss may be due to irreversible adsorption to the membrane. With acylase, an enzyme with a molar mass of $\sim$90,000 g mol^{-1}, retention is insufficient when a new Dow HFU-10 (cut-off 70,000 g mol^{-1}) hollow fiber unit is applied [46]. After 25 volume exchanges this loss ceases to be detectable. This pattern is found in quite a number of applications [86, 97, 106] and may be due to partial clogging of pores in the case of isotropic membranes (e.g. Dow hollow fiber) or to adsorption of catalyst to anisotropic ones, thereby blocking the passage for further molecules. This effect can be used to increase the retention capability of membranes significantly by "precoating" with high molar mass polymers as dextran [37] or PVA [86]. It is recommended to use membranes with a cut-off at least of a factor 3, better 5 less than the molar mass of the enzyme.

Reliable stability data can seldom be extracted from the literature for several reasons. Often the reactor is only operated for a few residence times, thus rendering an extrapolation rather speculative. Operating times less than 5 d should not be accepted for the evaluation of the operational stability further taking into account the actual mean residence time. Even the main parameters of reactor operation are not reported [88-90] or reported in a contradictory manner [95]. Thus, the catalyst mass referred space time cannot be calculated. Another common practice is to evaluate the stability at high substrate conversion [83, 84] without taking into account the highly nonlinear interrelation between conversion and the mass of active catalyst under such conditions. If there are no data available which allow the calculation of the deactivation constant (see Eq. (20)), the stability should be measured at low degrees of conversion where it can be assumed that the decline of conversion is not much lower than the activity decline. Beyond the statement of either practically no loss [92, 97, 100] or negligible loss [88] during reactor operation, only a few explicit data are given. Thus, for repetitive batch operation during lactose hydrolysis in whey permeate, a deactivation of 6.1% per cycle is reported [91]. A cephalosporin acetylesterase lost $\sim$2.5% of its activity per operating cycle [85]. For continuous operation with acylase a deactivation constant of 3.4% d^{-1} has been found compared to a storage stability of 2% d^{-1} [83] at 37 °C. Formate dehydrogenase deactivated at a rate of 0.83% d^{-1} [87].

Although the activity in membrane reactors should be utilizable quantitatively owing to the predominantly homogeneous distribution of catalysts, lower activity yields have been reported. During continuous operation, acylase could be utilized with an effectiveness of 60–70 % in a single UFMR and of 42 % in a 2-stage cascade with catalyst recycling [83]. This difference is due to variation of the modules applied. For invertase, an effectiveness of 45 % was obtained [93]. Differences between theory and experiment have also been reported by Hong et al. [84] although not expressed numerically for normal operation. They found that concentration polarization has a negative effect on catalyst effectiveness. If the reactor was operated without stirring, the activity declined to 42 % of the expected value. When the gel layer had been redissolved, an effectiveness close to unity was attained. The interpretation of high flux experiments by Hong et al. [84] might be somewhat speculative. They found that cellobiase activity declined very fast when the reactor was operated at high flux, while normal speed stirring was applied. Lowering the flux, and thereby the concentration polarization, had no effect on the activity yield. They interpreted this phenomenon as shear deactivation near the membrane where the enzyme is expected to be concentrated in the laminar boundary layer. It is far more probable that a gel layer is not found under these conditions, but that the retention of the catalyst decreased, as has been reported by Butterworth et al. [97]. In the unstirred experiment, the gel layer formed may have prohibited an extensive loss of catalyst (see also Sect. 2.1). The interpretation could be verified by analyzing the protein content (active and denatured enzyme) of the reactor after operation.

Although flux is an important operating variable, its dependence is not often specified for operation with low molar mass substrates. Boudrant and Cheftel [92] found a steep decrease of flux when the invertase concentration exceeded 1.5 kg m^{-3}. For acylase in conjunction with symmetric membranes, it has been shown that the membrane resistance rises steeply at low concentrations as a function of enzyme concentration [46]. A further increase of enzyme concentration has only a slight influence.

The flux is a more limiting variable for membrane reactor operation with macromolecular substrates. Applications in this field are summarized in Table 7.

All experiments refer to the conversion of starch, cellulose or proteins and have been carried out on laboratory-scale. An interesting detail may be that only one [166] of these applications employs repetitive batch operation, although it might be the most favourable mode of operation (see also discussion in Sect. 3.2). The continuous mode of operation predominates. The diafiltration mode (the reactor is fed with substrate and the products are eluted with buffer) is employed to test for limit-substrate (which cannot be further cleaved by the enzyme) or to study the product composition as a function of time. Another reason is the aim to overcome product inhibition.

Butterworth et al. [97] studied the conversion of starch by α-amylase and glucoamylase. With a 0.4 % starch concentration, they attained steady state in a UFMR. After 8 volume replacements, the flux had already declined to 50 % of the initial value. Although enzymatically thinned 9 % starch solution has been employed in the case of glucoamylase, the flux dropped to 60 % during the first volume replacement. They assumed that no undegradable substrate had accumulated in the UFMR. If the steady state would have been achieved together with no accumulation of undegradable substances, the flux should also have attained a steady state. This was certainly not the case.

An empirical correlation for flux decline has been given by Azhar and Hamdy [98]

Table 7. Enzymes employed in membrane reactors to convert high molar mass substrates

Enzyme	Source	Native	Modified	Substrate	Repet. batch	Semi-contin.	Continuous	Δp const.	J const.	Membrane Flat	Membrane Tubular	Ref.
α-Amylase	*B. subtilis*	×		Starch			×	×		×		97)
α-Amylase	—	×		Starch			×		×	×		166)
β-Amylase	Sweet potato	×		Starch				×		T		98)
Cellulase	*T. viride*	×		Mun. sludge		D	F	×		×		99)
Cellulase	*T. viride*	×		Solka Floc		D	×	×		×		100)
Cellulase	*T. viride*	×		Solka Floc			×		×		×	101)
Glucoamylase	—	×		Starch			×		×	×		166)
α-Chymotrypsin	—	×		Whey		D			×	×		10)
α-Chymotrypsin	—		×	Casein			×		×	×		102)
Proteases	—	×		Cotton seed prot.		D		×		T		103)
Proteases	—	×		Proteins			×	×		×		104)
Proteases	—		×	Proteins		D	×	×		T		105)
Proteases	—	×		Casein, Albumin	×		×		×	×		166)
Proteases	—	×		Soy protein			×		×		×	167)
Trypsin	—	×		Fish protein		D	×	×		×		106, 107)

D diafiltration mode
F substrate fed discontinuously
T thin-channel system

for the degradation of starch by β-amylase. In this case, the reactor was coupled to a thin channel system.

Cinq-Mars and Howell [99] studied the treatment of raw primary settled municipal sludge with cellulase. 75% of the cellulosic material can be solubilized mainly as cellobiose. The semicontinuous reaction experiment is interesting in the sense that the gel-like substrate is changed to a slurry of fine particles. This product can be ultrafiltered without having any fear of complete clogging of the membrane. Ghose and Kostick [100] evaluated different membranes for a prior concentration of cellulase and operated a membrane reactor continuously for 2.5 volume replacements. They initiated the reaction by placing a 48 h predigest of a 20% cellulose in the stirred ultrafiltration cell and fed a 10% aqueous suspension of cellulose to yield a mean residence time of 3.3 h. Since highly converted substrate is replaced by new substrate of lower concentration, the rapid attainment of a constant product concentration is far from being understood. An exact mass balance for cellulose and the amount of enzyme in the reactor could have been helpful. Another membrane reactor has been operated in the batch reaction/batch diafiltration mode by changeing the volumes each cycle time. Diafiltration has been carried out by feeding enzyme solution. Each cycle, fresh substrate was added. A yield of 60.7% is obtained after 10 d operation. Unfortunately the enzyme concentration in the diafiltration feed was not reported. The flux in the case of a cellulose digest solution corresponds to approximately 2–6% of the water flux for different membranes.

Henley et al. [101] operated two kinds of membrane reactors continuously and compared the results to those obtained in a normal CSTR for the hydrolysis of cellulose. The installations used consisted of a well mixed stage in series with a mixed ultrafiltration cell or a capillary membrane module, respectively. They found a considerable increase in performance for the membrane reactors, especially for that equipped with the capillary module. They interpret their findings as being due to a removal of inhibiting products. Although some operating parameters are omitted, it is unlikely that the results can be interpreted in this way. Some batch experiments and the CSTR operation have been carried out in the same reactor vessel under the same concentration conditions. A substrate conversion of 62% was found in the CSTR for a mean residence time of 12.25 min while this conversion was attained in the batch reactor only after 30 min.

Product inhibition cannot be overcome in membrane reactors behaving approximately like a CSTR by eluting the product while feeding new substrate, because these reactors operate at outlet conditions. The only way to increase their performance is to concentrate the substrate in the reactor. Thus, it works at a higher substrate concentration level than the concentration in the feed. When the conversion is defined with respect to the feed substrate concentration, very high conversions will be obtained, depending on whether or not a steady state has already been achieved (conversion = 100%). Since the conversion in all three cases of continuous reactor operation is significantly higher than it would be predicted by the batch measurements, retention of substrate particles must be assumed even for the CSTR operation, as a result of sedimentation of the cellulose.

The enzymatic hydrolysis of proteins seems to be of important industrial interest since the supplementation of beverages with soluble proteins is of particular interest. For the formulation of transparent liquid food, the proteins have to be cleaved into polypeptides to increase their solubility at normal pH and to remain soluble even after heat treatment. Although proteases are cheap enzymes and the hydrolysis poses, in principle, no problems, one disappointing fact has prohibited the large scale application of such products: in the course of the hydrolysis, oligopeptides of bitter taste are produced [110]. Roozen and Pilnik [104] reported that this characteristic might be due to peptides of a molar mass around $1,000 \text{ g mol}^{-1}$. Thus the idea of using membrane reactors to washout the oligopeptides of higher size immediately after formation and thus preventing further hydrolysis [110] has evolved. It does not seem that the problem of bitterness has been overcome [104, 105], because: proteases may act rather statistically and the aimed wash-out of appropriate peptides would only occur during operation in the uneconomic diafiltration mode and not in the fully continuous mode of operation (permanent feed of proteins) as a result of the behaviour of the reactors employed (CSTR). As a result of the broad residence time distribution of the molecules in the CSTR, further depolymerization occurs.

Although it has been found by Roozen and Pilnik [104] that products from a diafiltration experiment exhibited a 'bland' taste and that by feeding proteins continuously, it changes to 'bitter and beany' taste, this finding has not been further examined.

Another problem of using proteases in their native state is their autolysis [102, 105, 106], which requires a constant and considerable feeding of proteases during continuous reactor operation [106] or a stabilization by reticulation to yield essentially soluble catalyst preparations. Although the stabilization obtained is quite pronounced [107, 105], the proteolytic activity drops steeply due to the copolymerization and low temperatures have to be maintained during continuous reactor operation [102]. Continuous experiments lasting 150 h (24 volume replacements) employing 15% casein solution as substrate have been reported [105]. They yield 48% final activity. It seems that proteins are rarely completely hydrolysed [168]. This leads to a gradual increase in substrate concentration in the reactor which affects the achievable flux significantly [103, 104, 106, 107, 166]. Pressurization/depressurization cycles have been employed [107] to enhance the flux. Cunningham et al. [103] employed a dialysis reactor to overcome the problem of membrane plugging but did not report the recovery of the enzyme from unreacted substrate.

As has been discussed in Sect. 3.2, the porous wall tubular reactor (PWTR) can be applied to convert macromolecular substrates. Since the problems involved are similar to those discussed for UFMR, only the literature, which can be divided into theoretical treatments [62, 111, 112] and experimental findings [112-115] need be cited.

Studies of pressure driven operation with enzymes covalently linked or adsorbed to ultrafiltration membranes have been directed towards fouling phenomena. The coupling of proteases and cellulases to the membrane surface can decrease membrane fouling. Thus proteases coupled to membranes significantly decrease the decline of flux during protein processing [116-119]. This technique can obviously not be applied for enzyme processing.

Another technique is to activate a membrane by functional reagents and then force an enzyme solution through it in order to covalently bind the enzyme onto the membrane and into the pores as reported by Gregor [120, 121]. Substrate solution is forced through the membrane and the reaction takes place only at and within the membrane. This technique is interpreted in comparison to the case where mass transfer is achieved by diffusion. If the pores are equally accessible to the substrate, the productivity should be higher in any case since diffusional restriction can be avoided. The activity yields reported by Gregor are 60 to 75%, although the method employed does not give the exact yield and depends on the reaction conditions. The excessive yield for *Escherichia coli* lactase might reflect not only the selective coupling of this enzyme but rather a nonlinearity of its activity with respect to its concentration. Simon et al. [122, 123] used a similar strategy to prepare active membranes. They forced enzyme solution through microporous membranes followed by a crosslinking of the enzyme in the pores. They then compared the activity obtained by enforced flow operation to that obtained by diffusive operation. This comparison is sometimes misleading for technical applications since it is undertaken at very low substrate concentrations. When the maximum concentration gradient is limited in this way, it is not surprising that the pressure driven operation is found to achieve a far higher productivity [123]. The pressure driven operation is the more appropriate way to use the enzymes economically, provided that the enzyme in the membrane is equally accessible and that it is not washed out. Schmidt-Kastner [124] reports a 'half-life' for subtilisin employed in this way of 125 d for the optical resolution of N-acylmethionine via the methyl ester. This specification is derived from an extrapolation of the conversion decay at substrate conversions around 96 to 94%. This reactor configuration is of interest when extremely short residence times are desired. This is the case for reaction systems where autohydrolysis of the substrate or product affects the yield or the purity of the product. The hydrolysis of penicillins, cephalosporins and the optical resolution of amino acids via the methyl esters are such reaction systems.

A special kind of ultrafiltration membrane reactor has been studied which is characterized by gelification of enzymes on the surface of ultrafiltration membranes. Since this technique is closely related to normal UFMR operation except that the enzyme is not kept homogeneously distributed, only the literature is cited [125–140].

Although the concentration polarization is maximal when this technique is applied, it may be favourably employed where short residence times are required or when the catalyst is significantly stabilized.

5.2 Coenzyme-dependent Enzyme Systems

Since coenzymes act as transport metabolites between different enzymes, they have to be employed in a soluble state to achieve maximum activity. In the biological environment, coenzymes are continually re-used, because the utilization by one species of enzyme regenerates it for another. Coenzyme-dependent enzymes are interesting for industrial processes since they catalyse oxido-reductive, group transfer and synthetic reactions. However, coenzymes are rather expensive substances, are normally utilized in a mol per mol product basis and have a high molar mass (e.g. NAD 663 g mol^{-1}, ATP 605 g mol^{-1}). For industrial use, they have to be regenerated and re-used. In consequence, the enzyme which catalyzes the desired reaction has to be either combined will another enzyme whose reaction regenerates the coenzyme, or the coenzyme has to be regenerated by chemical means.

An example of such a regeneration cycle is given in Fig. 29. The desired reaction is catalyzed by alanine dehydrogenase, yielding L-alanine from pyruvic

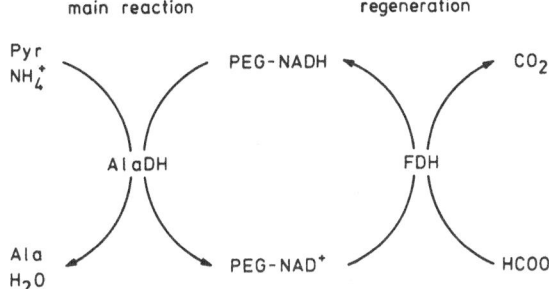

Fig. 29. Example for a coenzyme-dependent enzyme system with cofactor regeneration

acid and ammonium ions, and requires NADH as coenzyme. The oxidized coenzyme NAD^+ can be reduced for further utilization, for example, by formate dehydrogenase which oxidizes formic acid to carbon dioxide by reducing the coenzyme to NADH.

The entire catalytic system has to be immobilized for continuous re-use. Since the coenzyme has to retain its access to both enzymes, membrane reactors can be advantageously employed because the reactants therein can be kept in solution and homogeneously distributed. The main question is how to retain the coenzyme by means of ultrafiltration membranes, since their molar mass is of the same order of magnitude of the products in question. As has been pointed out for enzymes, coenzymes can also be linked to soluble polymers. NAD has been intensively studied from this point of view. It has been covalently bound to soluble poly-ethylene glycol (PEG) [141–143], sepharose [144], polylysine [145], alginic acid [146], vinyl-derived polymers [147–149], dextran [144, 150], polyethyleneimine [151] and agarose [152]. ADP/ATP has been bound to dextran [153]. These artificially enlarged coenzymes can be used for continuous processing.

Table 8. Coenzyme-dependent enzyme systems employed in membrane reactors

Enzymes		Coenzyme		Reaction		Ref.
Main	Regenerating	Type	Polymer	Substrate	Product	
ADH	Ferricyanide	NAD	Dextran Sepharose	Ethanol	Acet-aldehyde	[144]
LDH	ADH	NAD	Polylysine	Pyruvate	Lactate	[145]
LDH	ADH	NAD	PEG	Pyruvate	Lactate	[141]
LDH	FDH	NAD	PEG	Phenyl-pyruvate	D-Phenyl-lactate	[154]
AlaDH	FDH	NAD	PEG	Pyruvate	L-Alanine	[154]
LeuDH	FDH	NAD	PEG	α-Keto-iso-caproate	L-Leucine	[155]
Hexokinase	Acetate kinase	ADP/ATP	Dextran	Glucose	Glucose 6-phosphate	[153]

ADH	alcohol dehydrogenase	AlaDH	alanine dehydrogenase
FDH	formate dehydrogenase	LDH	lactate dehydrogenase
LeuDH	leucine dehydrogenase	PEG	polyethylene glycol

Table 8 summarizes those reaction systems which have been studied in membrane reactors. Most of these studies are directed towards the properties of the enlarged coenzyme. Therefore, common enzymes like LDH and alcohol dehydrogenase (ADH) as well as hexokinase and acetate kinase have been employed to show the feasibility of this concept. One of the main problems is the choice and development of an adequate regenerating system. The substrate for the regeneration cycle has to be inexpensive and the resulting product must not affect the desired reaction. The use of ADH might, therefore, be limited to low conversions since its product, acetaldehyde, is known to de-stabilize the activity of enzymes due to its reactivity with aminogroups. The regenerating system which seems to best meet the needs for processes requiring NADH is formate dehydrogenase (FDH). It utilizes formate as substrate and produces carbon dioxide, a harmless product. The degassing of CO_2 has only to be considered at acidic pH, low pressures and high conversion, and it can be technically overcome. The relatively low mass specific activity of FDH is no problem, especially when larger reactors are employed. The modification of NAD normally leads to a considerable increase in its K_m value although the maximum rate of reaction is sometimes maintained. Since coenzymes naturally exhibit low K_m values (e.g. 0.02 mol m^{-3}) an increase of one order of magnitude results in technically reasonable concentrations for enlarged coenzymes. The processes could be significantly improved, if the natural instability of coenzymes could be overcome. The NAD turnover number of 18,200 mol substrate converted per mol NAD used reported for the AlaDH/FDH system [154] shows that coenzyme-dependent enzyme systems may soon become economically feasible. A general survey on the use of coenzymes is given by Wang and King [163].

5.3 Microorganisms

The application of ultrafiltration in combination with microorganisms is directed towards the use of whole cells and organelles for catalytic purposes, production of metabolites, continuous biomass production and sewage treatment for water purification and recycling.

Chromatophores from *Rhodospirillum rubrum* have been used by Pace et al. [156] to regenerate ATP from ADP continuously in a membrane reactor. When adenylate kinase is added to the photosynthetic regeneration system, AMP can also be re-converted to ATP. The deactivation is found to be approximately 0.04 and 0.03 h^{-1}, after a steep initial decrease of activity. Fumarate has to be added, which, as assumed, might shift the reaction equilibrium or affect an inhibitory compound formed during the reaction.

Whole cells of *Caldariella acidophila* have been entrapped in various membranes by Drioli et al. [157–159]. These membranes have been employed for pressure driven operation to hydrolyse p-nitrophenyl-β-D-galactopyranoside by means of the bacterial β-galactosidase. Wang et al. [160] have studied a semi-continuous 'membrane fermentor' for the production of a protease from *Clostridium histolyticum*. After 12 h of batch operation fresh substrate was fed continuously to the fermentor. Filtrate was withdrawn continuously by means of an ultrafiltration device which retained the cells and enzyme. A 2.5 fold increase in biomass was observed while the enzyme activity yield increased by a factor of 3.9 compared to batch operation. Since no operating variables were reported, the interpretation can only be speculative. It seems that the feeding of substrate in the late exponential phase may be solely responsible for the enhanced enzyme production since it leads to further cell growth. The interpretation often found in the literature, that the removal of toxic low molar mass substances by ultrafiltration might be of significant importance, is at least misleading since CSTR-type fermentors are employed (see discussion in Sect. 5.1). During batch processes, enzyme production is limited by the depletion of substrate. This is not the case during the semi-continuous operation of a membrane fermentor. An indication of the amount of enzyme produced per unit substrate consumed would be valuable for the interpretation of this promising procedure.

The main advantage of continuous membrane fermentors is that they can be operated at higher biomass concentration than in conventional fermentor operation. For this purpose a 'rotorfermentor' has been designed by Margaritis and Wilke [161]. Another important economic reason for employing ultrafiltration in combination with fermentation is the possibility of water recycling. This has been studied by Goto et al. [162] for the production of biomass from methanol by a *Methylomonas* sp. They found that at least 50% of the filtrate could be recycled.

Jeffries et al. [164] have studied the production of methane from glucose by methanogenic bacteria using a membrane fermentor operated in semi-continuous mode. It seems that this has been achieved by repetitive batch/feeding periods. Since it is reported that 85% of methane and CO_2 is evolved during the feeding/ultrafiltration period, a completely continuous operation would certainly enhance productivity. Such steady state experiments would also lead to a better understanding and interpretation of such results.

A special design of a semi-continuous membrane fermentor for sewage treatment has been proposed by Lambert and Jamet [165]. It is characterized by the use of the recycle flow through the ultrafiltration unit to feed oxygen and agitate the fermentor contents by means of an injector.

6 Concluding Remarks

During the last decade, ultrafiltration has become an established process in biotechnology. Although nearly every paper describing the purification of enzymes includes an ultrafiltration step, its description is seldom adequate nor is it in any way optimized. Despite numerous examples, information beyond yield and improvement in specific activity can normally not be extracted from the literature. The wide acceptance of ultrafiltration as a unit operation in laboratory- as well as industrial-scale systems indicates that reliable, non-fouling, high flux membranes have become commercially available. The design of ultrafiltration modules and units has advanced, so that ultrafiltration can now be employed on practically any scale. The ever present effect of concentration polarization can be described in most cases in terms of the gel polarization model. Since ultrafiltration in biotechnology seldom deals with simple systems, a reliable layout can only be accomplished by means of experimental data. This implies that deviations from commonly used models might be taken into account by introducing empirical extensions.

   Since the layout for concentration processes is largely predetermined by the solution which has to be processed, the optimization of diafiltration has to take into account that a dilute solution leads to a higher flux but requires a larger eluant volume. The optimum depends on economic considerations. Concentration processes are normally carried-out as batch operations with complete recycling of the solution, to increase the solute concentration as slowly as possible, at constant transmembrane pressure. Continuous ultrafiltration is only applied for large scale processing, which requires a stagewise design of the ultrafiltration plant. Diafiltration is usually carried out in either 'closed loop' or by cyclic batch operation of dilution and concentration steps. Fractionation of enzymes by ultrafiltration seems to yield satisfactory results only if either the difference in molar mass is at least one order of magnitude or if it can be combined with specific binding to other macromolecules. For concentration and diafiltration of enzymes in complex mixtures it may often be favourable to use membranes giving non-quantitative retention, since the simultaneous removal of impurities compensates for small losses of the product.

   However, for the application of biocatalysts in membrane reactors it is absolutely necessary to achieve quantitative retention. The review of the literature dealing with membrane reactors has shown that this area is rather unclear despite numerous examples and their application on industrial-scale [11]. It has been necessary to revise the theory of shear deactivation in laminar flow and a new theoretical guideline had to be formulated for the layout of membrane reactors. One of the most obstinate beliefs found in the literature is that a main advantage of membrane reactors would be the removal of inhibitors and toxins of low molar mass 'as they are formed'. Substances which are not retained at all by the membrane are governed by the same balance equations as in any normal reactor of comparative design. Only substances which are essentially retained by the membrane can influence performance when compared to conventional reactors. If substrates of high molar mass are applied, it has to be assured that a steady state can be achieved under the given conditions. It has to be considered that at steady state, the substrate concentration in the reactor can be much higher than in the feed and that the feed mass flow must be equal to the exit mass flow of substrate and product.

   While continuous processing of high molar mass substrates seems to have not yet attained a state of technical feasibility, there is an enormous potential for the conversion of low molar mass substrates in membrane reactors. Even the use of coenzyme-dependent enzyme systems seems likely to become feasible beyond laboratory-scale in the near future. The concept of membrane fermentors still needs further development before its obvious advantages might be exploited.

7 Nomenclature

Symbols and commonly used units

a	$m^2\ m^{-3}$	volume specific area
A	m^2	area
c	$mol\ m^{-3}$	concentration
D	$m^2\ s^{-1}$	diffusion coefficient

d	m	diameter
E	kg m^{-3}	enzyme concentration
E$_a$	J mol^{-1}	energy of activation
f	—	ratio
J	m s^{-1}	flux (area specific volumetric flow rate)
J$_i$	mol m^{-2} s^{-1}	molar flux of substance i
k$_2$	mol kg^{-1} s^{-1}	reaction rate constant
k$_d$	m s^{-1}	mass-transfer coefficient
k$_{de}$	s^{-1}	deactivation rate constant
K$_{ic}$	mol m^{-3}	constant for competitive product inhibition
K$_m$	mol m^{-3}	Michaelis-Menten constant
l	m	length
m	kg	mass
M	kg kmol^{-1}	molar mass
n	—	number of stage(s)
p	Pa = N m^{-2}	pressure
r	m	radius
r$_m$	mol kg^{-1} s^{-1}	catalyst mass specific reaction rate
r̄$_m$	mol kg^{-1} s^{-1}	catalyst mass specific productivity
r$_v$	mol m^{-3} s^{-1}	reactor volume specific reaction rate
R	—	apparent retention
R$_m$	—	intrinsic retention
S	mol m^{-3}	substrate concentration
t	s	time
t̄	s	mean residence time
u	m s^{-1}	linear velocity
V	m^3	volume
V̇	m^3 s^{-1}	volumetric flow rate
W	Pa s m^{-1}	hydraulic resistance
X	—	substrate conversion
x	m	local coordinate

Greek Symbols

γ	s^{-1}	shear rate (absolute value)
δ	m	laminar boundary layer thickness
ε	—	porosity
η	Pa s	dynamic viscosity
ν	m^2 s^{-1}	kinematic viscosity
π	Pa	osmotic pressure
ϱ	kg m^{-3}	density
σ$_p^2$	m^2	variance of pore diameter
σ$_i^2$	m^2	variance of particle diameter
τ	Pa	shear stress
τ$_m$	kg s m^{-3}	catalyst mass referred space time
θ	—	dimensionless time (t/t̄)

Indices

b	bulk phase
c	convective
d	diffusive — (excl. k_d)
e	end-, final-
E	enzyme-
f	filtrate-
g	gel-
h	hydraulic-
i	referred to species i
m	membrane- (excl. τ_m, r_m, K_m)
n	referred to stage n, number of stages
o	initial-, feed-
p	pore-
P	productivity referred-
r	radial-
R	reactor-
s	saturation-
S	separation unit-
t	tube-
V	volume
w	water (eluant)-
max	maximum-
opt	optimum-
rec	recycle-
Σ	total

Abbreviations

ADH	alcohol dehydrogenase
A(D)TP	adenosine (di)triphosphate
AlaDH	alanine dehydrogenase
CMR	cascade of completely equipped UFMR
CRMR	cascade recycle membrane reactor
CSTR	continuous stirred tank reactor (ideal)
FDH	formate dehydrogenase
LDH	lactate dehydrogenase
LeuDH	leucine dehydrogenase
MR	membrane reactor (general)
PEG	polyethylene glycol
PFTR	plug flow tubular reactor
PVA	polyvinyl alcohol
PWTR	porous wall tubular reactor
SBR	stirred batch reactor
TRMR	tubular recycle membrane reactor
UFMR	continuously operated single stage ultrafiltration membrane reactor

8 References

1. Michaels, A. S., in: Polymer Science and Technology, Vol. 13, Ultrafiltration Membranes and Applications, (Cooper, A. R. ed.), p. 1, Plenum Press, New York 1980
2. Sourirajan, S., Loeb, S.: Advan. Chem. Ser. *38*, 117 (1962)
3. Michaels, A. S.: CHEMTECH, p. 36 Jan. 1981
4. Madsen, R. F.: Hyperfiltration and Ultrafiltration in Plate-and-Frame Systems, Elsevier Publ. Co., 1977
5. Strathmann, H.: Chem. Tech. *7* (8), 333 (1978)
6. Porter, M. C.: Ind. Eng. Chem. Prod. Res. Develop. *11* (3), 234 (1972)
7. Rauch, K.: Doctoral thesis, TH Aachen, W.-Germany 1978
8. Blatt, W. F., Feinberg, M. P., Hopfenberg, H. B.: Science *160*, 224 (1965)
9. Michaels, A. S., in: Progress in Separation and Purification, Vol. 1, p. 297, Wiley & Sons, 1968
10. Blatt, W. F., Hudson, B. G., Robinson, S. M., Zipilivan, E. M.: Anal. Biochem. *22*, 161 (1968)
11. BMFT-Mitteilungen *4*, p. 60, Bonn: Bundesministerium für Forschung und Technologie 1982
12. Graetz, L.: Ann. Physik. *25*, 337 (1885)
13. Lévêque, J.: Ann. Mines, *14*, 201, 305, 381 (1928)
14. Gröber, H., Erk, S., Grigull, V.: Fundamentals of Heat Transfer, p. 233, McGraw-Hill, New York 1961
15. Dittus, F. W., Boelter, L. M.: Univ. Calif., Berkeley, Publ. Eng. *2*, 443 (1930)
16. Harriott, P., Hamilton, R. M.: Chem. Eng. Sci. *20*, 1073 (1965)
17. Blatt, W. F., Dravid, A., Michaels, A. S., Nelson, L., in: Membrane Science and Technology, Flinn, (J. E. ed.), p. 47, Plenum Press, New York—London 1970
18. Pitera, E. W., Middleman, S.: Ind. Eng. Chem., Process Des. Develop. *12* (1), 52 (1973)
19. Copas, A. L., Middlemann, S.: ibid. *14* (2), 143 (1974)
20. Shen, J. J. S., Probstein, R. F.: ibid. *18* (3), 547 (1979)
21. Probstein, R. F., Shen, J. S., Leung, W. F.: Desalination *24*, 1 (1978)
22. Van der Waal, M. J., Van der Velden, P. M., Koning, J., Smolders, C. A., van Swaay, W. P. M.: Desalination *22*, 465 (1977)
23. Bixler, H. J., Rappe, G. C.: US 3,541,006, Nov. 17, 1970
24. Ebner, H., Enenkel, A.: US 3,974,068, Aug. 10, 1976
25. Ebner, H.: Chem.-Ing.-Tech. *53*, 25 (1981)
26. Brown, C. E., Tulin, M. P., Van Dyke, P.: Chem. Eng. Progr., Symp. Series, Vol. 67 (114), p. 174, 1971
27. Holtz, H. W.: Doctoral thesis, TH Aachen, W.-Germany 1979
28. Rautenbach, R., Holtz, H.: Chem.-Ing.-Tech. *51* (12), 1241 (1979)
29. Leung, W.-F., Probstein, R. F.: Ind. Eng. Chem. Fund. *18* (3), 274 (1979)
30. Goldsmith, R. L., de Filippi, R. P., Hossain, S., Timmins, R. S., in: Membrane Processes in Industry and Biomedicine, (Bier, M. ed.), p. 267, Plenum Press, New York—London 1971
31. Yan, S. H., Hill, C. G., Amundson, C. H.: J. Dairy Sci. *62* (1), 23 (1979)
32. Baker, R. W., Strathmann, H.: J. Appl. Polymer Sci. *14*, 1197 (1970)
33. Breslau, B. R., Kilcullen, B. M.: Enzyme Eng. *3*, 179 (1975)
34. Flaschel, E., Wandrey, C., in: Characterization of Immobilized Biocatalysts, DECHEMA Monogr., Vol. 84, p. 337, Verlag Chemie, Weinheim 1979
35. Lonsdale, H. K., in: Industrial Processing with Membranes, (Lacey, R. E., Loeb, S. eds.), chapter VIII, p. 123, Wiley Interscience, New York 1972
36. de Filippi, R. P.: Chem. Proc. Eng. N.Y. *10*, 475 (1977)
37. Flaschel, E.: Enzyme Eng. *6*, 57 (1980)
38. Timmins, R. S., in: Freeze Drying and Advanced Food Techn., (Goldblith, S. A., Rey, L., Rothmayr, W. W. eds.), chapter 31, p. 503, Academic Press, London 1975
39. Michaels, A. S.: Sep. Sci. Technol. *15* (6), 1305 (1980)
40. Jandel, L., Schulte, B., Bückmann, A. F., Wandrey, C.: J. Membrane Sci. *7*, 185 (1980)
41. Butterworth, T. A., Wang, D. I. C.: Ferment. Technol. Today, Proc. IV IFS, p. 195, 1972
42. Blatt, W. F.: J. Agr. Food Chem. *19* (4), 589 (1971)
43. Hatch, R. T., Price, J. D.: AIChE Symp. Ser. *74*, No. 172, 226 (1978)
44. Tutunjian, R. S., Reti, A. R.: AIChE. Symp. Ser. *74*, No. 178, 210 (1978)
45. Melling, J.: Process Biochem., p. 7, Sept. 1974

46. Wandrey, C., Flaschel, E., Schügerl, K.: Ger. Chem. Eng. *1*, 39 (1978)
47. Charm, S. E., Wong, B. L.: Biotech. Bioeng. *12*, 1103 (1970)
48. Charm, S. E., Wong, B. L.: Biorheology *12*, 275 (1975)
49. Tirrell, M., Middleman, S.: Biotechnol. Bioeng. *17*, 299 (1975)
50. Reese, E. T., Ryu, D. Y.: Enzyme Microb. Technol. *2*, 239 (1980)
51. Charm, S. E., Wong, B. L.: Science *170*, 466 (1970)
52. Charm, S. E., Wong, B. L.: Biotech. Bioeng. *20*, 451 (1978)
53. Charm, S. E., Matteo, C. C.: Meth. Enzymol. *22*, 476 (1971)
54. Charm, S. E., Lai, C. J.: Biotech. Bioeng. *13*, 185 (1971)
55. Virkar, P. D., Narendranathan, T. J., Hoare, M., Dunnill, P.: ibid. *23*, 425 (1981)
56. Tirrell, M., Middleman, S.: ibid. *20*, 605 (1978)
57. Tirrell, M., Middleman, S.: Biophys. J. *23*, 121 (1978)
58. Tirrell, M.: J. Bioeng. *2*, 183–193 (1978)
59. Charm, S. E., Wong, B. L.: Enzyme Microb. Technol. *3*, 111 (April 1980)
60. Thomas, C. R., Dunnill, P.: Biotech. Bioeng. *21*, 2279 (1979)
61. Thomas, C. R., Nienow, A. W., Dunnill, P.: ibid. *21*, 2263 (1979)
62. Shah, Y. T., Remmen, T.: J. Heat Mass Transfer *14*, 2209 (1971)
63. Neubeck, C. E.: US 4,233,405, Nov. 11, 1980
64. Gerstenberg, H., Sittig, W., Zepf, K.: Chem.-Ing.-Tech. *52* (1), 19 (1980)
65. Wang, D. I. C., Sonoyama, T., Mateles, R. I.: Anal. Biochem., *26*, 277 (1968)
66. Mercer, J. E., in: Proc. Int. Workshop on Techn. for Protein Separation and Improvement of Blood Plasma Fractionation, (H. E. Sandberg ed.), No (NIH) 78–1422, p. 160, Washington DC: U.S. Govt. Printing Office 1978
67. Ng, P., Lundblad, J., Mitra, G.: Separation Sci. *11* (5), 499 (1976)
68. Pace, G. W., Schovin, M. J., Archer, M. C.: Separation Sci. *11* (1), 65 (1976)
69. Howell, J. A., Velicangil, O., Le, M. S., Herrara Zeppelin, A. L.: Ann. Acad. Sci., N.Y. *369*, 355 (1981)
70. Heinen, W., Lauwers, A. M.: Arch. Microbiol. *106*, 201 (1975)
71. Nielsen, W. K., in: Rotenburger Symp. 1978, Fermentationtechnik, p. 195, 1978
72. Martinache, L., Henon, M. P., in: Methods of Plasma Protein Fractionation, (Curling, J. M. ed.), p. 223, Acad. Press, London 1980
73. Barbaric, S., Kozulic, B., Ries, B., Mildner, P.: Biochem. Biophys. Res. Commun. *95* (1), 404 (1980)
74. Hummel, W., Schütte, H., Kula, M.-R.: Eur. J. Appl. Microbiol. Biotechnol. *12*, 22 (1981)
75. Andersson, R. E.: Biotechnol. Letters *2* (5), 247 (1980)
76. Wang, D. I. C., Sinskey, A. J., Sonoyama, T.: Biotech. Bioeng. *11*, 987 (1969)
77. Balfanz, T., Hicke, H.-G., Paul, D., Schwarz, H.-H.: Chem. Techn. *31* (11), 557 (1979)
78. Kroner, K. H., Hustedt, H., Kula, M.-R.: unpublished
79. Trinel, P. A., Leclerc, H.: Ann. Microbiol. *127B*, 201 (176)
80. Dion, P., Goulet, J., Lachance, R. A.: J. Inst. Can. Sci. Technol. Aliment. *11* (2), 78 (1978)
81. Bartling, G. J., Barker, C. W.: Biotech. Bioeng. *18*, 1023 (1976)
82. Chambers, R. P., Cohen, W., Baricos, W. H.: Meth. Enzymol. *44*, 291 (1976)
83. Wandrey, C., Flaschel, E.: Adv. Biochem. Eng. *12*, 147 (1979)
84. Hong, J., Tsao, G. T., Wankat, P. C.: Biotechnol. Bioeng. *23*, 1501 (1981)
85. Abbott, B. J., Cerimele, B., Fukuda, D. S.: ibid. *18*, 1033 (1976)
86. Weiss, R.: Doctorjal thesis, TU Hannover, W.-Germany 1978
87. Wichmann, R.: Doctoral thesis, TU Clausthal, W.-Germany 1981
88. Roger, L., Thapon, J. L., Maubois, J. L., Brule, G.: Le Lait *551–552*, 56 (1976)
89. Roger, L., Maubois, J. L., Thapon, J. L., Brule, G.: Ann. Nutr. Alim. *32*, 657 (1978)
90. Roger, L., Thapon, J. L., Brule, G., Maubois, J. L.: nordeuropaeisk mejeri-tidsskrift *77* (1–2), 38 (1977)
91. Norman, B. E., Severinsen, S. G., Nielsen, T., Wagner, J.: The World Galaxy for the World Dairy Industry, *7*, 20 (1978)
92. Boudrant, J., Cheftel, C.: Biochimie *55*, 413 (1973)
93. Bowski, L., Shah, P. M., Ryu, D. Y., Vieth, W. R.: Biotech. Bioeng. Symp. *3*, 229 (1972)
94. Bowski, L., Ryu, D. Y.: Biotechnol. Bioeng. *16*, 697 (1974)

95. Ryu, D. Y., Bruno, C. F., Lee, B. K., Venkatasubramanian, K.: Ferment. Technol. Today, Proc. IV IFS, p. 307 (1972)
96. Cawthorne, M. A.: DOS 2,356,630, Int. Cl.: C 07 d, 99/16, 22. Mai 1974
97. Butterworth, T. A., Wang, D. I. C., Sinskey, A. J.: Biotech. Bioeng. 12, 615 (1970)
98. Azhar, A., Hamdy, M. K.: ibid. 23, 1297 (1981)
99. Cinq-Mars, G. V., Howell, J.: ibid. 19, 377 (1977)
100. Ghose, T. K., Kostick, J. A.: ibid. 12, 921 (1970)
101. Henley, R. G., Yang, R. Y. K., Greenfield, P. F.: Enzyme Microb. Technol. 2, 206 (1980)
102. O'Neill, S. P., Wykes, J. R., Dunnill, P., Lilly, M. D.: Biotech. Bioeng. 13, 319 (1971)
103. Cunningham, S. D., Cater, C. M., Mattil, K. F.: J. Food Sci. 43 (5), 1477 (1978)
104. Roozen, J. P., Pilnik, W.: Enzyme Microb. Technol. 1, 122 (1979)
105. Boudrant, J., Cheftel, C.: Biotech. Bioeng. 18, 1735 (1976)
106. Bhumiratana, S., Hill, C. G., Amundson, C. H.: J. Food. Sci. 42 (4), 1016 (1977)
107. Payne, R. E., Hill, C. G., Amundson, C. H.: ibid. 43, 385 (1978)
108. Marshall, J. J., Rabinowitz, M. L.: Biotech. Bioeng. 18, 1325 (1976)
109. Wykes, J. R., Dunnill, P., Lilly, M. D.: Biochim. Biophys. Acta 250, 522 (1971)
110. Pilnik, W.: Gordian 75 (5), 208 (1973)
111. Closset, G. P., Shah, Y. T., Cobb, J. T.: Biotech. Bioeng. 15, 441 (1973)
112. Katoh, S., Yanagida, T., Sada, E.: J. Chem. Eng. Japan 11 (2), 143 (1978)
113. Closset, G. P., Cobb, J. T., Shah, Y. T.: Biotech. Bioeng. 16, 345 (1974)
114. Tachauer, E., Cobb, J. T., Shah, Y. T.: ibid. 16, 545 (1974)
115. Madgavkar, A. M., Shah, Y. T., Cobb, J. T.: ibid. 19, 1719 (1977)
116. Howell, J. A., Knapp, J. S., Velicangil, O.: Enzyme Eng. 4, 267 (1978)
117. Jenq, C. Y., Wang, S. S., Davidson, B.: Enzyme Microb. Technol. 2, 145 (April 1980)
118. Velicangil, O., Howell, J. A.: Biotech. Bioeng. 19, 1891 (1977)
119. Velicangil, O., Howell, J. A.: ibid. 23, 843 (1981)
120. Gregor, H. P.: US 4,033,822, July 5, 1977
121. Gregor, H. P.: DOS 2,650,920, Int. Cl. C 12 D 13/00, 18. Mai 1977
122. Simon, S., Bloch, R., Caplan, S. R.: Biotechnol. Appl. Protein Enzymes, p. 169 (1977)
123. Bloch, R., Caplan, R. S., Simon, S.: DOS 2,553,649, Int. Cl. C 07 B 29/00, 10. Jun. 1976
124. Schmidt-Kastner, G., in: Bioreaktoren, 2. BMFT — Statusseminar „Bioverfahrenstechnik" Jülich, 1979, p. 63, Bonn: Bundesministerium für Forschung und Technologie 1979
125. Cantarella, M., Remy, M. H., Scardi, V.: Chem. Eng. Sci. 34, 1213 (1979)
126. Cantarella, M., Gianfreda, L., Palescandolo, R., Scardi, V., Greco, G., Alfani, F., Iorio, G.: J. Solic-Phase Biochem. 2 (2), 163 (1977)
127. Cantarella, M., Remy, M.-H., Scardi, V., Alfani, F., Iorio, G., Greco, G.: Biochem. J. 179, 15 (1979)
128. Copobianco, G., Drioli, E., Ragosta, G.: J. Solid Phase Biochem. 2 (4), 315 (1977)
129. Drioli, E., Bellucci, F.: Desalination 26, 17 (1978)
130. Drioli, E., Scardi, V.: J. Membr. Sci. 1, 237 (1976)
131. Drioli, E., Mendia, J., Molinari, R.: Desalination 24, 193 (1978)
132. Gianfreda, L., Greco, G.: Biotechnol. Letters 3 (1), 33 (1981)
133. Greco, G., Alfani, F., Cantarella, M., Gianfreda, L., Palescandolo, R., Scardi, V.: Chem. Eng. Commun. 7, 145 (1980)
134. Greco, G., Albanesi, D., Cantarella, M., Gianfreda, L., Palescandolo, R., Scardi, V.: Eur. J. Appl. Microbiol. Biotechnol. 8, 249 (1979)
135. Greco, G., Albanesi, D., Cantarella, M., Scardi, V.: Biotech. Bioeng. 22, 215 (1980)
136. Greco, G., Alfani, F., Iorio, G., Cantarella, M., Formisano, A., Gianfreda, L., Palescandolo, R., Scardi, V.: ibid. 21, 1421 (1979)
137. Scardi, V., Cantarella, M., Gianfreda, L., Palescandolo, R., Alfani, F., Greco, G.: Biochimie 62, 635 (1980)
138. Maculan, T. P., Hourigan, J. A., Rand, A. G.: J. Dairy Sci. 61 (suppl. 1), 114 (1978)
139. Drioli, E., Gianfreda, L., Palescandolo, R., Scardi, V.: Biotech. Bioeng. 17, 1365 (1975)
140. Vorsilak, P., McCoy, B. J., Merson, R. L.: J. Food Sci. 40, 431 (1975)
141. Furukawa, S., Katayama, N., Iizuka, T., Urabe, I., Okada, H.: FEBS Letters, 121 (2), 239 (1980)
142. Bueckmann, A. F., Morr, M., Johansson, G.: Makromol. Chem. (in press)
143. Bueckmann, A. F., Kula, M.-R., Wichmann, R., Wandrey, C.: J. Appl. Biochem. (in press)

144. Malinauskas, A. A., Kulis, J. J.: Appl. Biochem. Microbiol. *14* (6), 706 (1978)
145. Yamazaki, Y., Maeda, H., Suzuki, H.: Biotech. Bioeng. *18*, 1761 (1976)
146. Coughlin, R. W., Aizawa, M., Charles, M.: ibid. *18*, 199 (1976)
147. Furukawa, S., Urabe, I., Okada, H.: Eur. J. Biochem. *114*, 101 (1981)
148. Furukawa, S., Sugimoto, Y., Urabe, I., Okada, H.: Biochimie *62*, 629 (1980)
149. Muramatsu, M., Urabe, I., Yamada, Y., Okada, H.: Eur. J. Biochem. *80*, 111 (1977)
150. Weibel, M. K., Fuller, C. W., Stadel, J. M., Bückmann, A. F. E. P., Doyle, T., Bright, H. J.: Enzyme Eng. *2*, 203 (1974)
151. Wykes, J. R., Dunnill, P., Lilly, M. D.: Biochim. Biophys. Acta *286*, 260 (1972)
152. Wykes, J. R., Dunnill, P., Lilly, M. D.: Biotech. Bioeng. *17*, 51 (1975)
153. Yamazaki, Y., Maeda, H., Suzuki, H.: Eur. J. Biochem. *77*, 511 (1977)
154. Wandrey, C., Wichmann, R., Bueckmann, A. F., Kula, M.-R.: Enzyme Eng. *6*, 453 (1980)
155. Wichmann, R., Wandrey, C., Bueckmann, A. F., Kula, M.-R.: Biotech. Bioeng. *23* (12) (1981)
156. Pace, G. W., Yang, H. S., Tannenbaum, S. R., Archer, M. C.: ibid. *18*, 1413 (1976)
157. De Rosa, M., Gambacorta, A., Esposito, E., Drioli, E., Gaeta, S.: Biochimie *62*, 517 (1980)
158. Drioli, E., Iorio, G., Molinari, R., De Rosa, M., Gambacorta, A., Esposito, E.: Biotech. Bioeng. *23*, 221 (1981)
159. Drioli, E., Gaeta, S., Carfagna, C., De Rosa, M., Gambacorta, A., Nicolaus, B.: J. Membrane Sci. *6*, 345 (1980)
160. Wang, D. I. C., Sinskey, A. J., Butterworth, T. A., in: Membrane Science and Technology, (Flinn. E. ed.), p. 99, Plenum Press, New York 1970
161. Margaritis, A., Wilke, C. R.: Devel. Industrial Microbiol. *14*, 159 (1972)
162. Goto, S., Kuwajima, T., Okamoto, R., Inui, T.: J. Ferment. Technol. *57* (1), 47 (1979)
163. Wang, S. S., King, C.-K.: Adv. Biochem. Eng. *12*, 119 (1979)
164. Jeffries, T. W., Omstead, D. R., Cardenas, R. R., Gregor, H. P.: Biotech. Bioeng. Symp. *8*, 37 (1979)
165. Lambert, S., Jamet, B.: Fr. 2,340,451-C 16, 3. Jul. 1980
166. Walon, R. G. P.: Dos 2,039,222, Int. Cl. C 12 d, 13/10, 25. Feb. 1971; Fr. 2,056,692, 18 Jun. 1971
167. Deeslie, W. D., Cheryan, M.: J. Food Sci. *46*, 1035 (1981)
168. Cheryan, M., Deeslie, W. D., in: Polymer Science and Technology, Vol. 13, Ultrafiltration Membranes and Applications, (Cooper, A. R. ed.), p. 591, New York: Plenum Press, New York 1980

Molecular Cloning in Heterologous Systems

Karl Esser and Christine Lang-Hinrichs
Lehrstuhl für Allgemeine Botanik, Ruhr-Universität, D-4630 Bochum 1, FRG

During the last years the knowledge accumulated by fundamental research of DNA recombination in vitro became more relevant to biotechnology.

One of the main advantages of this so-called genetic engineering is the possibility to recombine genetic material from organisms which cannot be hybridized by means of the classical techniques of chromosomal recombination.

After a brief survey of the historical background of DNA transformation, the principal requirements of transforming DNA molecules are outlined, and widely used DNA vehicles are compiled.

Strategies and limits of molecular cloning are commented prior to the discussion of the main host organisms for foreign DNA.

In this review we have tried to select from the enormous number of results those which are of biotechnological relevance at present and those which may have future perspective in applied research.

1 Introduction

Recombination, one of the essential parameters of evolution, is responsible for the exchange of genetic material between individuals of the same species. In eukaryotes recombination is achieved predominantly as a result of sexual propagation (meiotic recombination) and rather infrequently during the course of vegetative propagation in parasexual processes (mitotic recombination). In prokaryotes recombination is initiated exclusively by various mechanisms of parasexual processes like bacterial conjugation, transduction, transformation.

The investigation of the different mechanisms of recombination which was especially successful in the case of fungi, bacteria and bacteriophages [1-3], has not only been important for fundamental research but also for applied research, e.g. plant and animal breeding, and strain improvement in biotechnology [4, 5].

However, the great handicap of the classical recombination procedures is that the exchange of genetic material involves homologous systems. Only in rare exceptions is a transgression of the species limits possible, due to experimental tricks (e.g. [6]).

To obtain heterologous recombination is especially important for breeding purposes and for biotechnology. From the fundamental studies of vegetative hybridization of animal cells [7] it became evident that the species limit is not a definitive barrier to the coexistence of heterologous genetic material. However, the break-through has been achieved only by molecular cloning[1] [10]. This technique allows the transfer, with the assistance of a specific vector, of distinct DNA sequences into host cells irrespective of whether there is a relationship between donor and host or not, e.g. from eukaryotes to prokaryotes. Certainly, this technique which opened the violently developing field of new biology or new genetics, has also its limits. This is due to the incompatibility and instability between the two heterologous DNAs when present in a common physiological machinery [11-14].

An advantage of molecular cloning is that, in contrast to classical recombination genetics, short DNA segments and even single genes may be transferred due to specific selection systems. This possibility of selected recombination is important for biotechnology if there is a chance to obtain both a suitable vector and a suitable host for replication and expression of the heterologous genome fragment.

This review is addressed to persons dealing with biotechnology. After a brief introduction of the problems of molecular cloning, some of the results obtained up to date with the new genetics in heterologous cloning will be critically discussed with respect to their application in biotechnology. (For general information see Refs. [15-25]).

2 Principles

Molecular cloning is based on the introduction of DNA molecules into cells, a process called *"genetic transformation"*.

---

[1] The concept *clone* is a rather old term. It was introduced by Webber [8] to describe "a population of cells or organisms derived from a single cell or common ancestor by mitosis". In this context it seems necessary to mention the term "nuclear cloning" which describes the "surgical placement of nuclei from accessible tissues into an unfertilized egg" [9]

Genetic transformation has to be clearly distinguished from "malignant transformation" which means the induction of tumourous growth in cells infected with oncogenic viruses. In this review the term "transformation" will be used to signify genetic transformation and will include the term "transduction" — the specialized mode of genetic transformation via virus genomes.

The first transformation was reported already in 1928 when Griffith [26] succeeded in transferring genetic information between different pneumococcal strains. The significance of these experiments, however, was not understood until Avery and coworkers [27] identified DNA as the transforming principle. Subsequently, several groups reported transformation in bacteria [28, 29].

Various efforts were also made to obtain transformation in eukaryotes after treatment with foreign DNA.

Especially Benoit and coworkers [30] caused a sensation by reporting morphological inheritable changes in ducks postnatally injected with DNA originating from different races. Similar transformations were performed on *Drosophila* eggs [31] and several plant species [32]. But these results were difficult to explain, in part not even reproducible [33]. Thus, it could not be unequivocally shown that foreign DNA was present and persisted in the host cell or the host organism. More substantial indications of transformation were obtained from experiments with *Neurospora crassa*. Mishra and Tatum [34] and Mishra et al. [35] found an increased rate of reversion of marker genes in auxotrophic mutants after treatment with wild type DNA. Some of these revertants transmitted their genotype to the sexual progeny.

However, significant progress was not made till experimental conditions were essentially altered and the new techniques of in vitro recombination of DNA were applied. A successful transformation requires the following criteria:

1) The transforming DNA consists of defined segments.

The prerequisite to this criterion comes from the work of Arber [36, 37] who demonstrated that specific endonucleases in bacteria are able to split double-stranded DNA at sites having a specific nucleotide sequence. It was Berg [38] and later on Cohen et al. [10] who recognized the significance of these restriction enzymes to produce distinct DNA fragments. Arber and Berg were awarded the Nobel Prize for their contribution in 1978 and 1980, respectively.

2) Organisms which serve as hosts for the transformation of DNA show a rather low degree of differentiation.

The first experiments were performed with unicellular organisms, such as the prokaryotic bacteria *Escherichia coli* and *Bacillus subtilis*. Later on it was possible to include eukaryotes like the yeast *Saccharomyces cerevisiae*, plant protoplasts, animal cells and even multicellular organisms like the hyphal fungi *Podospora anserina* and *Neurospora crassa*.

3) As a result of transformation cloned DNA sequences can be identified biochemically in the cells in addition to their associated phenotypically recognizable selective markers.

For this purpose the technique of radioactive DNA-DNA-hybridization developed by Southern [39] ("Southern blotting technique") is frequently used.

These three criteria are in general sufficient to perform transformation within species limits. However, hybridization exceeding these limits in both prokaryotes and eukaryotes requires the integration of the DNA to be transferred into a transport vehicle. This *vector* consists of DNA compatible with the host and must be able to replicate and express in the host.

In general the term vector is used in biology to characterize a vehicle which transports information such as organisms transferring diseases (e.g. mosquitos). In molecular biology especially in molecular cloning the term vector is used in a more specific sense. It comprises a molecule which is suited for the transport of genetic information. In most cases DNA originating from bacterial plasmids or viral genomes are used as vectors of this kind.

Only after having developed appropriate vectors, heterologous molecular cloning became a successful technique. Therefore, the next chapter will be devoted to a more detailed description of the various vector systems.

3 Vectors

3.1 General Characterization

Any vector suitable for molecular cloning needs to accomplish the following criteria:

3.1.1 Compatibility with the Host Cell

It has to be guaranteed that after entering into the cell the vector becomes a stable part of the host cell genome. However, some time ago, it became evident that DNA originating from different organisms even within one species or race may be incompatible when brought together in one cell or into cellular contact. The phenomenon, called *heterogenic incompatibility*, is very basic in biology and ranges from tissue incompatibility (organ transplantation in mammals, graft incompatibility in higher plants) via cellular incompatibility (e.g. heterokaryon incompatibility in fungi) to molecular incompatibility in prokaryotes (e.g. failure of transformation in bacterial) (for additional literature see [40]).

Especially, the latter phenomenon that DNA molecules when transferred from one bacterial cell to another are rapidly restricted by endonucleases of the host, is of particular relevance to molecular cloning. Only the fact that in some cases the bacterial cell may loose by mutation its capacity to restrict foreign DNA allows the development of molecular cloning (literature in Ref. [15]).

Thus, at present, the *E. coli* strains used for transformation experiments are restriction negatives derived from *E. coli* K 12 C 600. Therefore, in this bacterium the effects of heterogenic incompatibility at the level of DNA are overcome.

This is not the case in *Bacillus subtilis*. This bacterium is frequently used as host for molecular cloning, because it does not produce toxins like *E. coli*. Despite the fact that many successful cloning experiments have been performed in *B. subtilis*, the genetic and biochemical basis for the instability of hybrid DNA is not yet understood (Ref. in [41]).

The phenomenon of DNA incompatibility is found to be very common in bacteria (for Refs. see [42, 147]). It also occurs in the eukaryotic *S. cerevisiae* during the experiments to establish it as host for molecular cloning [148].

3.1.2 Autonomous Replication in the Host Cell

After being accepted by the host cell, the multiplication of the vector DNA must be guaranteed. The vector has to include a region that may induce and control replication in the host cell. Such a DNA sequence representing a unit of autonomous replication is named *replicon* [43] and carries the determinants for the enzymes of initiation and the origin of replication. Replicons are present on every autonomous DNA molecule. This is definitely the case in *plasmids*, detected in numerous prokaryotic organisms.

Plasmids are covalently closed circular extrachromosomal genetic elements whose replication and segregation to daughter cells at divisions is independent of the actual genome of the cell [44].

The prerequisite is fulfilled by the vectors of viral origin, such as DNA of the bacteriophage lambda [45] or the mammalian simian virus (SV 40) [46].

In this context it seems necessary to stress that the source of replicons is not restricted to the "classical" plasmids or lambda DNA. Especially, if molecular cloning is extended to eukaryotic hosts, appropriate compatible replicons have to be available. Only a very few plasmids of eukaryotic origin are known so far. The best analyzed example is the 2 μm plasmid of the yeast *Saccharomyces cerevisiae* [47, 48], which was already successfully used as cloning vehicle in eukaryotes ([49], see p. 21). Whereas the yeast plasmid is believed to be of cytoplasmic origin [50, 51], there are other plasmids or plasmid structures originating from the mitochondrial DNA, such as the pl DNA of the filamentous fungus *Podospora anserina* (see also Table 1) [52].

The techniques of molecular cloning have also provided the methods for isolating and employing as vectors replicons from more complex structures, such as chromosomes or mitochondrial DNA.

The first successful experiments using chromosomal origins of replication were reported for the yeast *S. cerevisiae* [63, 64]. Recently, data revealed that also a fragment of the mitochondrial genome of *Podospora anserina* may be used as replicon for cloning in this organism (see also Table 2) [65].

This perspective opens new pathways in molecular cloning of heterologous DNA, because it seems to be possible to cleave chromosomal DNA or rather (less complex) mt DNA from any eukaryote in order to construct its specific vector.

3.1.3 Possession of a Selective Marker

In order to identify after transformation and replication the very few transformed cells (e.g. in bacteria 10^{-2}–10^{-4}) the vector must carry a marker gene. Fortunately, bacterial plasmids first used as vectors usually carried several genes responsible for resistance against antibiotics. This was taken advantage of, because it is easy to select resistant bacteria. Furthermore, it was known that the reversion rate of plasmid genes is rather low as compared to chromosomal genes.

The use of resistance genes as phenotypic markers has a further advantage since it is possible to splice a resistance gene which after integration of foreign DNA is inactivated (= insertional inactivation). The rediscovery of this hybrid vector is only possible when at the same time at least one more marker gene is present on the same vector. The principle of this most commonly used method is described in Fig. 1 [19, 20].

3.2 Compilation of Commonly Used Vectors

At present, a great variety of vectors are used for various purposes in different organisms, some of them are commercially available. The majority of these vectors were constructed by using naturally occurring plasmids (mostly of bacterial origin) in which parts of prokaryotic DNA and, to a lesser extent, DNA from eukaryotes (e.g. yeasts and filamentous fungi) were integrated. In some cases, the basic part of the vector may be viral DNA too.

Table 1. Compilation of circular DNA molecules in eukaryotes (except normal mitochondrial and chloroplast DNA)

Organism	Phenotype	Specific designation	Size of monomer	Origin	Ref.
1. Nuclei-associated					
Saccharomyces cerevisiae	Wild	2 μm-DNA	6.3 kb	Unknown	53)
Schizosaccharomyces pombe		2 μm/3 μm	6.3/9.5 kb		54, 55)
2. Mitochondria-associated					
a. mt DNA-sequences					
Saccharomyces cerevisiae	Rho-mutants	—	varies	mt DNA, deletion	56)
Neurospora crassa	Poky-mutants	—			57)
Neurospora crassa	Stopper-mutants				58)
Podospora anserina	Wild-type strains	pl DNA	2.4 kb	mt DNA, excision	52)
Aspergillus amstelodami	Ragged-mutants	pl DNA	0.8 kb		59)
Zea mays	Male sterile mutants	pl DNA	varies		60, 61)
b. mtDNA-independent					
Neurospora crassa	Wild	Plasmid 2225	3.6 kb	Unknown	62)
			4.2 kg		
		—	5.0 kb		

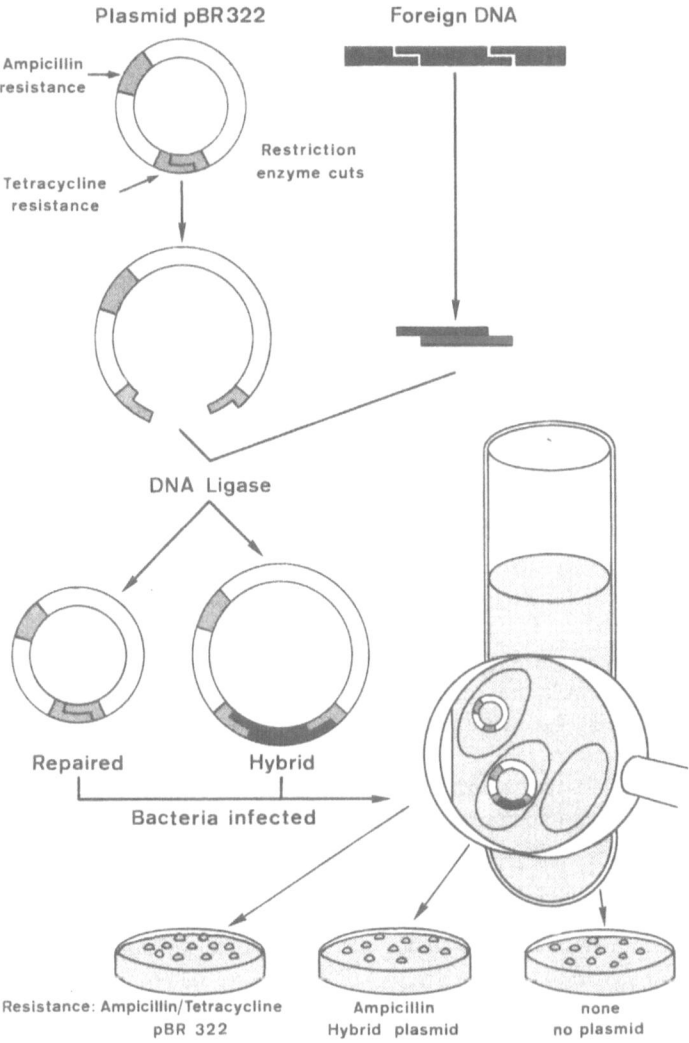

Plasmid pBR322   Foreign DNA

Ampicillin resistance

Restriction enzyme cuts

Tetracycline resistance

DNA Ligase

Repaired   Hybrid

Bacteria infected

Resistance: Ampicillin/Tetracycline   Ampicillin   none
pBR 322   Hybrid plasmid   no plasmid

Fig. 1. Scheme of direct molecular cloning. The bacterial plasmid pBR 322 (white) carrying two marker genes (hatched) (see also Table 1) is used as a vector. By addition of a specific restriction endonuclease pBR 322 is cleaved within the marker gene causing tetracycline resistance. The foreign DNA (black) treated with the same endonuclease is then added. After ligation two types of plasmids are found in the mixture: the repaired original plasmid and a newly built hybrid plasmid containing a part of the foreign DNA. After addition to a bacterial population and subsequent plating on different media, three types of bacterial colonies may be identified according to their marker genes. It thus becomes evident that, due to insertional inactivation of the tetracycline gene, only the bacteria which gained ampicillin resistance carry the hybrid plasmid and therewith the foreign DNA (after [66])

Any compilation of these vectors suffers from the handicap that it is sometimes hard to decipher the organism from which the vector DNA is derived. This situation is certainly not relieved by the designation of the vectors since there is no commonly adopted nomenclature, yet. Therefore, in the list of Table 2 we refrain from presenting the source of the vectors and prefer to describe the organisms in which the vectors are able to replicate and express. Other compilations of vectors may be found in Refs. [15, 19, 20, 23].

Vectors of viral origin frequently mentioned and employed are those based on bacteriophage lambda. Its well studied genome was one of the first DNAs used in molecular cloning [90].

Bacteriophage lambda is a temperent phage, i.e. it can be integrated both into the genome of the host cell and replicate along with it once per generation and, upon induction, replicate outside the host genome and produce viral progeny. Its genome was shown to contain regions dispensible with the lytic growth cycle which, for molecular cloning experiments, may be replaced by foreign DNA segments (replacement vectors).

Lambda vectors used today (see Table 2) contain several mutations (restricting growth to a specific host strain) and recognition sites for commonly used restriction enzymes (facilitating the integration of foreign DNA segments).

A detailed review of the characteristics and potentials of lambda vectors is given by Williams and Blattner [91].

To achieve efficient infection and replication of recombinant phage genomes the overall length of the molecule has to be within certain limits because only DNA molecules of approximately the same length as native lambda DNA are efficiently packed into the phage heads.

It could also be shown [92] that apart from the length requirements there is only a small part of the lambda genome necessary for the packaging procedure in vitro. This part consists of the so-called *cos* site and a small region in its proximity.

Lambda DNA is a double stranded molecule of about 15 μm in length. It exists as linear molecule inside the phage particle and possesses complementary single stranded termini at either side (cohesive ends) which associate to form the *cos* site upon injection of the DNA into the host cell.

Based on the knowledge of the packaging system of phage lambda both in vivo and in vitro a new class of cloning vectors was developed, the *cosmids* which combine the cos site of lambda and a common resistant plasmid [93].

Cosmids infect the host cell like lambda phages and behave like a plasmid in the cell (i.e. replicate without subsequently lysing the cell). A main advantage of these vectors is the possibility to clone rather large DNA fragments (up to about 14 μm) whereas DNA of this length cloned in plasmid vectors is not always kept stable [15].

Another phage of *E. coli* has recently been developed as a cloning vector, i.e. the single-stranded DNA phage M13 [94].

Infection of a cell by M13 does not lead to lysis of the cell, but cell growth is retarded and phage particles are continuously produced and excreted. Packaging of the single-stranded and circular DNA is independent of the length of the molecule so that foreign DNA may be integrated without impeding the viability of the phage. Manipulations like restriction and ligation are performed with the double-stranded replicative form of the viral DNA.

M13 vectors usually employed now possess a reliable phenotypical marker like the one listed in Table 2.

Table 2. Compilation of vectors used for molecular cloning in prokaryotes and eukaryotes

Host organism	Designation	Contour length (μm)	Marker genes and characteristics	Ref.
Escherichia coli	pSC101	2.8	Tet[r]	67)
and other gram-negative bacteria	ColE1	2.0	Colicin[imm], Colicin[prod]	68)
	*pMB9	1.8	Colicin[imm], Tet[r]	69)
	pBR313	2.7	Colicin[imm], Amp[r], Tet[r]	70)
	*pBR322	1.4	Amp[r], Tet[r]	71)
	*pBR325	1.7	Amp[r], Tet[r], Cm[r]	72)
	pACYC177	1.2	Amp[r], Kan[r]	73)
	pACYC184	1.3	Tet[r], Cm[r]	73)
	*λgtWES:λB	13.0	Critical size of molecule Bacteriophage lambda derived	74)
	*M13mp7		Lac complementation Phage M13 derived	75)
	*pHC79	2.0	Amp[r], Tet[r], lambda cos-site	76)
Bacillus subtilis	pBC16	1.4	Tet[r]	77)
and other gram-positive bacteria	pC194	1.0	Cm[r]	78)
	*pUB110	1.4	Kan[r]	79)
	pE194	1.2	Ery[r]	80)
	pBS161-1	1.2	Tet[r]	81)
Escherichia coli +	pJK3	2.4	Tet[r], Amp[r]	81)
Bacillus subtilis	pCS194	2.4	Amp[r], Cm[r]	82)
	pCD1	2.6	Tet[r], thy[+]	83)
Escherichia coli +	pJDB219	4.0	Tet[r], leu[+]	49)
Saccharomyces cerevisiae	pJDB248	4.4	Tet[r], leu[+]	49)
	pMP78	2.8	Amp[r], Cm[r], leu[+]	84)
	pLC544	3.3	Amp[r], Tet[r], trp[+]	85)
	Yep6	2.5	Amp[r], his[+]	13)
	Yep7	1.8	Amp[r], Tet[r], trp[+]	13)
	Yep13	3.4	Amp[r], Tet[r], leu[+]	86)
E. coly + Herpes simplex-DNA	*HSV-106	2.5	Amp[r], tk[+]	87)
E. coli + *Podospora anserina*	pSP4	2.1	Amp[r], Senescence induction in P.a.	89)
	pKP402	3.0		89)

The first column specifies the organisms in which the vectors replicate and express. Vectors which may be used in two different organisms are called "hybrid vectors" or "hybrid plasmids" or sometimes "chimaeric plasmids". They carry two different replicons and marker genes that may be selected in either system.

The designation of vectors as given in the second column either refers to the author's name (e.g. pSC101 stands for plasmid Stanley Cohen) or to the cloning organism (e.g. Yep13 stands for yeast plasmid).

The size of the vectors is given by their contour length (in μm). In other publications the molecular weight (in M Dalton) or the number of nucleotides (in kilobase pairs) is used. (1 μm = 2.07 MD = 3.14 kb).

Marker genes are abbreviated as follows: ampicillin resistance Amp[r], chloramphenicol resistance Cm[r], tetracycline resistance Tet[r], kanamycin resistance Kan[r], erythromycin resistance Ery[r], colicin immunity Col[imm], colicin production Col[prod], thymidine prototrophy thy[+], leucine prototrophy leu[+], tryptophane prototrophy trp[+], histidine prototrophy his[+], thymidinekinase activity tk[+].

Commercially available vectors are marked by an asterisk.

They are predominantly used in molecular cloning experiments for subsequent sequencing analyses [95].

4 Strategies

Before detailing the various strategies currently applied in molecular cloning experiments, the main purpose and potential of these techniques need to be summarized:
a) Analysis of genome and gene structure.

Molecular cloning makes DNA segments available in extremely concentrated and pure quality for further analysis, e.g. for sequencing. For this purpose correct replication of the desired genome segment in the heterologous system (in this case almost exclusively in *E. coli*) is sufficient. Correct transcription and translation of a potential gene need not to be ensured.

b) Synthesis of industrially important proteins.

The genetic information for proteins of biotechnological importance is mostly obtained from eukaryotes whereas the most common host organisms suitable for practical applications are mainly microorganisms like *E. coli*, *B. subtilis* or *S. cerevisiae*.

c) Manipulation of plants for breeding purposes.

Molecular cloning in plants is mainly developed for solving problems such as biological nitrogen fixation which will be discussed in Chapter 4.

The various strategies applied to reach the different goals are fundamentally based on common techniques which may be divided into two main groups.

4.1 Direct Cloning

The general principle of this strategy (illustrated in Fig. 1) is that the segment of DNA to be cloned is inserted into a vector and transferred to the host cell. On the assumption that the inserted DNA is compatible with the heterologous host, stable replication of the hybrid plasmid should take place.

The expression of the foreign DNA, however, is not always achieved by this technique which was originally developed for DNA multiplication [58]. Failure of expression will mainly be due to the inability of the host cell to cope with the regulating signals of the foreign gene(s), such as spliced sequences of eukaryotic mosaic genes[2]. In order to overcome this handicap a second but more time-consuming strategy was developed.

---

[2] The knowledge of molecular structure of genes originates almost exclusively from the study of prokaryotic and viral DNAs [96]. Quite recently, however, it became evident that this concept of gene structure is not in general valid for eukaryotic genes [97]. There are many examples showing that the information of an eukaryotic gene is not necessarily continuous (see Table 3). In contrast, coding regions (exons) alternate with non-coding regions (introns) [98]. These *mosaic genes* are also known as *spliced genes*.

Table 3. Compilation of some eukaryotic mosaic genes

Gene	Organism	Ref.
Actin	*Saccharomyces cerevisiae*	99)
	Sea urchin	100)
	Drosophila	101)
Alcohol dehydrogenase	*Drosophila*	102)
α-Amylase	Mouse	103)
Cytochrome b	*S. cerevisiae*	104)
Dihydrofolate reductase	Mouse	105)
Fibroin	Silk worm	106)
β-Globin	Rabbit	107)
	Mouse	108)
	Human	109)
Immunoglobulin	Mouse	110)
Insulin	Rat	111)
	Human	112)
Leghemoglobin	Soya-bean	113)
Lysozyme	Chicken	114)
Ovalbumin	Chicken	97)
Ovomucoid	Chicken	115)
Phaseolin	French bean	116)
Vitellogenin	*Xenopus*	117)

4.2 Indirect Cloning

The general principle of this strategy is that the segment of DNA to be cloned is modified in vitro prior to linkage to a vector. This technique was developed to overcome the difficulties which especially arose in the cloning of spliced eukaryotic genes in heterologous hosts. As illustrated in Fig. 2 (upper part), it is based on the discovery that in vivo the transcription of these mosaic genes is different from that of unspliced genes in that it comprises at least two steps: Firstly, the whole gene region (a) is, as in unspliced genes, transcribed into RNA (b). Secondly, the introns are cleaved from the transcript in one or more steps and the exons only are ligated to the mRNA (c) which now carries all genetic information necessary for the synthesis of the gene product [118]. In the cell this mRNA is the substrate for translation.

With this knowledge in mind a new strategy was developed to overcome the difficulties of discontinuous or spliced genes by way of indirect cloning. It was then only necessary to isolate mRNA from the donor cells instead of DNA [119, 120].

This mRNA is not used directly for cloning experiments as shown in Fig. 2 (lower part), but it needs to be transcribed into DNA before being inserted into a DNA vector [121].

By using the enzyme reverse transcriptase it is possible to synthesize a complementary single-stranded DNA (sometimes also called copy DNA, or briefly cDNA) from the mRNA template with a supplementary loop at its 3'-end consisting of a few nucleotides (d, e).

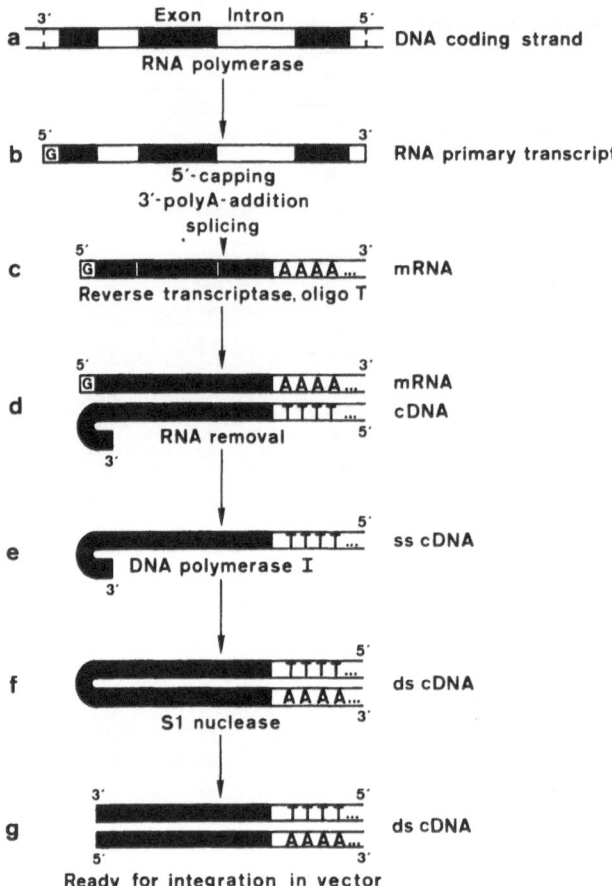

Fig. 2. Scheme of a strategy for indirect molecular cloning of a eukaryotic mosaic gene using the cDNA technique. For further explications see text

Due to subsequent action of DNA polymerase I, which uses the loop as a primer, a double-stranded cDNA is produced (f). Finally, the loop is cleaved by S1 nuclease. The double-stranded cDNA is now ready for insertion into an appropriate vector (g).

5 Limits

If foreign DNA is transferred to a host cell by a vector following either the direct or indirect strategy, there may occur problems at the various steps leading to the expression of the heterologous DNA:

1) *Coexistence of heterologous DNA in the host cell.* An incompatibility between the DNA of the host and the introduced DNA may be easily overcome (see also p. 146) by choosing a host organism which is devoid of a DNA restriction system [122].
2) *Replication.* In order to ensure replication of the heterologous DNA, the vector

carrying this DNA species must possess a replicon homologous to the host DNA and thus compatible with the cell system (see also p. 146).

3) *Transcription*. Promoters which, as binding sites for the RNA polymerase, are responsible for initiation of transcription, are specific to either prokaryotic or eukaryotic genes. Therefore, it must be ensured that an appropriate promoter is located on the vector in front of the gene to be expressed.

A typical *prokaryotic* promoter is composed of two specific sequences in front of the gene. Approximately 10 nucleotides apart from the transcribed sequence there is a characteristic nucleotide sequence called the "Pribnow box" (TATPuATG) [123, 124], and a second region involved in the initiation of transcription is a further 25 nucleotides away from the gene.
Transcription of *eukaryotic* genes is achieved by different types of RNA polymerase (I, II and III). RNA-polymerase II which transcribes the structural genes recognizes a region about 30 nucleotides in front of the point of initiation which was named TATA-Box (because of its characteristic nucleotide sequence of TATAAAA) or "Goldberg/Hogness-box" [125, 126]. Similar to promoters in prokaryotes there seems to be a specific region further away from the gene as well [127, 128].

Strategies to overcome this problem were developed almost exclusively for the cloning of heterologous (eukaryotic) genes in *E. coli*. The most commonly applied strategy is to join the heterologous DNA fragment to homologous promoters so that the RNA polymerase starts transcription at the homologous recognition site and reads through to transcribe the foreign DNA as well, under the assumption that there is no termination codon present in front of the foreign gene (e.g. [129]).

4) *Translation*. For this procedure one is faced with the fact that ribosomes of different structure (70 S and 80 S respectively) are involved in protein synthesis in prokaryotes and eukaryotes. From this follows that prokaryotic and eukaryotic genes have different recognition sequences (= ribosome binding sites).

In *prokaryotes* effective initiation is dependent on the interaction of the small ribosomal subunit (more specifically its 16S rRNA) and the messenger RNA. The recognition sequence is rich in purine nucleotides (adenine and guanine) and is located a few nucleotides in front of the gene sequence. Base-pairing between the 16S rRNA and this so-called "Shine-Dalgarno-sequence" (characteristically AGGA) provides the specificity of initiation [130]. mRNAs of *eukaryotes*, however, are devoid of such recognition sequences. Eukaryotic 80S ribosomes rather recognize a modification of the 5'-terminus of the mRNA (5'-cap which is a 7-methylguanosin) [131, 132]. Efficient translation of mRNAs in the eukaryotic system is also dependent on the 3' modification (poly-A tail) which seems to stabilize the mRNA molecule in the cell [133–135].

The most commonly applied strategy to render a eukaryotic gene translatable in a prokaryotic host is to fuse it with a homologous gene (see p. 159) [136]. However, it is also possible to fuse it with isolated homologous translation signal sequences (ribosome binding sites) [137] or, which is a more elaborate approach, to synthesize chemically the proper sequences or the whole gene [138, 139].

The chemical synthesis of a gene or of gene sequences is a further possibility of indirect cloning which is, however, feasible only for relatively short polypeptides (see p. 159). It has several advantages as compared to the cloning with cDNA, because it allows cloning to be restricted to those nucleotides that carry information for the gene product; moreover, regulation sequences may be adapted to the host system (for reviews see [140, 141]).

Another difficulty may arise from the recent discovery that the genetic code is not completely universal, since it could be shown that the genetic code used in mitochondria differs in a number of ways from the standard code [142].

This problem again may be overcome by using a chemically synthesized gene sequence which takes into account potential discrepancies.

The result of translation is the desired polypeptide which has to be traced among the cellular proteins. There are in principal two methods for detecting the expression of a cloned gene.

a) The synthesis of an active enzyme which restores some function in the host cell is followed, a method applicable e.g. to genes coding for metabolites cloned in auxotrophic host strains [143].

b) The peptide product is made out by testing its immunological [144] or its biological [145] activity.

This second approach is certainly more widely applicable since recognition of the "cloned" protein is independent of the existence of a homologous gene function in the host but also possible when gene expression is only partially correct [146].

5) *Coexistence of heterologous proteins in the host cell.* It is well-known that foreign proteins will provoke a defense reaction of the host cell comparable to the above mentioned DNA incompatibility.

There is no prediction possible which particular protein will be attacked by proteinases of the host. However, experience gained in cloning experiments with *E. coli* has shown that a degradation of this kind is rather common [145]. The heterologous protein may be protected against decomposition when its genetic information is integrated into a homologous gene, thus coding for a host-compatible fusion protein [129, 139].

6) *Contamination.* In any case the "cloned" protein has to be separated from the host cell. If a host like *E. coli* produces toxic substances this may cause problems [147]. Even in the most simplest case, when the cloned protein is excreted, the possibility of contamination with a toxic substance excreted as well, may not be overruled. The situation becomes more complex if the host cell has to be destroyed in order to gain the protein. Apart from the possibility that during cell lysis already degradation of the cloned protein may be initiated, the protein has subsequently to be separated by elaborate biochemical procedures from the cell debris. The only possibility to avoid these difficulties is to choose a host organism (like e.g. *Bacillus subtilis*) which does not produce toxins (see p. 161).

7) *Stability of the transformed organism.* Even if all criteria mentioned either do not apply or may be somehow overcome, there is no guarantee that the transformed host strain will remain stable during a prolonged asexual (or sexual) propagation (e.g. [149, 150]). One has to keep in mind that in contrast to chromosomes, extra-chromosomal DNA does not necessarily follow a programmed replication cycle [151]. This is especially true for plasmids [152]. There are some examples demonstrating that, especially in prokaryotes, DNA is lost during the course of various cell divisions if there is no selective pressure to keep it.

The same phenomenon is often found after heterologous cloning where foreign gene sequences are excised from the vector, e.g. in *B. subtilis* [147]. There is however no prediction so far how to overcome this handicap, but individual studies for each particular case are required.

6 Applications

In the previous chapters we have deliberately discussed the fundamentals of heterologous cloning in some detail. Thus, when presenting in this chapter a survey of examples of heterologous cloning, we are able to give a critical evaluation of the problems and achievements connected with these experiments. From the numerous data now available in this field we shall mainly deal with those concerning biotechnology, i.e. therapeutics and other substances of economic value. In other words we shall not discuss experimental results concerning fundamental problems such as regulation of gene expression or structural organization of genes and genomes. This section is divided according to the host organisms employed in molecular cloning experiments.

6.1 Prokaryotes

6.1.1 *Escherichia coli*

It was almost natural that *E. coli* was the first host organism to be used for cloning experiments as it is genetically the best studied prokaryote. Thus, the vectors first used for molecular cloning originated either from this gram-negative bacterium or were correlated with it (plasmids and DNA of bacteriophage lambda respectively) [10, 90]. Up to now there are numerous publications on successful heterologous cloning in *E. coli*.

In Table 4 some of the data are compiled under the following points of view:

First, the examples given for direct cloning are more or less complete. They have not much of a biotechnological relevance, though, and will not be discussed in detail. They are mainly given to show what kind of genes may be cloned directly and expressed in *E. coli*.

Whereas DNA of heterologous bacteria, no matter whether gramnegative or gram-positive, seems to be generally expressed, DNA of eukaryotic origin is not principally active in *E. coli*. Several eukaryotic genes involved in biosynthesis of metabolites (mostly of fungal origin as analyzed so far) may be expressed upon direct cloning. However, expression of many genes of higher eukaryotes is often rendered more difficult by their intron/exon structure and only possible via cDNA copies or chemically synthesized sequences.

Second, at present, there are numerous reports dealing with indirect cloning of eukaryotic DNA. Thus, we were obliged to confine ourselves to selected products which, on the one hand, seem to be most relevant to biotechnology and, on the other hand illustrate the relevance of some of the limits and strategies discussed above.

Insulin, a hormone responsible for the reduction of the blood glucose level may serve as an example that the same protein may be synthesized by the following different strategies. The first approach was to construct a cDNA copy (p. 153) of the proinsulin mRNA. In order to achieve efficient expression of the mammalian gene in the host, it was inserted into the coding sequence of the bacterial gene for β-galactosidase so that the homologous transcription and translation signals (see p. 155) would be used [136].

Insulin consists of two peptide molecules which are interconnected by two disulfide cross-linkages. The synthesis of insulin in the cell proceeds via a precursor called proinsulin from which the A- and B-chain are cleaved post-translationally.

Table 4. Examples of heterologous molecular cloning in *E. coli*

	Source of DNA	Gene or gene function	Ref.
Direct cloning			
	Bacillus subtilis	Thymidine metabolism	[11]
		Leucine metabolism	[153]
		Pyrimidine metabolism	[154]
	Bacillus cereus	Tetracyline resistance	[81]
	Bacillus licheniformis	Ampicillin resistance	[155]
	Staphylococcus aureus	Chloramphenicol resistance	[78]
		Tetracycline resistance	[156]
		Kanamycin resistance	[157]
		Streptomycin resistance	[157]
	Saccharomyces cerevisiae	his3, histidine metabolism	[143]
		leu2, leucine metabolism	[158]
		trp5, tryptophane metabolism	[159]
		argH, arginine metabolism	[160]
		arg4, arginine metabolism	[161]
		gal1, galactokinase	[162]
		trpl, tryptophane metabolism	[13]
		argE, arginine metabolism	[163]
		ura1, uracil metabolism	[164]
	Neurospora crassa	qa2, dehydroquinase	[165]
	Herpes simplex-virus	Thymidinekinase	[166]
Indirect cloning cDNA			
	Rat	Proinsulin	[136]
	Mouse	β-Endorphin	[145]
	Human	Growth hormone	[167]
		Interferon	[168]
		Urokinase	[169]
	Foot-and-mouth disease virus	Coat protein	[170]
	Hepatitis B virus	Core antigen	[171]
Chemical DNA synthesis			
	Bovine	Thymosin α_1	[172]
	Human	Somatostatin	[129]
		Insulin	[139]
		Interferon	[173]

The second approach to obtain insulin was made by means of a chemically synthesized DNA segment. The nucleotide sequence was derived from the known peptide sequence of the insulin molecule (with 51 amino acids altogether) and the information for the A- and B-chain was cloned in *E. coli* separately [139]. The purified peptide chains were joined in vitro to form the mature insulin molecule.

To ensure the expression of integrated foreign gene sequences they are frequently linked to defined promoters of genes of *E. coli* and coli phages [174]. Besides the β-galactosidase gene just mentioned more genes are used such as a gene of lambda bacteriophage or the β-lactamase gene of *E. coli* (present e.g. on plasmid vector pBR322) which were successfully employed for the synthesis of a protein of foot and mouth disease virus and the synthesis of urokinase, respectively [169,170].

In addition, the use of the β-lactamase gene, which is a secreted protein itself, will result (at least partially) in a transport of the hybrid protein into the periplasmic space [136].

In the majority of experiments the DNA sequences are integrated into the bacterial gene at an internal position which means that the required polypeptide has to be released after purification of the hybrid protein [129, 139]. Most simply, this is done in a chemical reaction using cyanogen bromide which specifically cleaves the polypeptide at a methionine residue, the amino acid that normally starts a polypeptide. This strategy has been used e.g. for the liberation of the insulin chains (discussed above) which do not possess internal methionine from the fusion protein.

Interferon, one of the most discussed therapeutically important proteins, is produced in certain mammalian cells in response to virus infections. Once present in an organism, it is believed to "*interfere*" with the propagation of viruses after superinfections immunizing the organism. Thus, the medical importance of interferon in the control of viral diseases or oncogenous cancer becomes obvious [175]. Despite the fact that both details and consequences of the therapeutical effects of interferon are not completely understood [176, 177], many efforts were undertaken to produce this protein.

The first method, i.e. isolation of interferon from stimulated cell cultures, has a rather low output and leads to a very expensive product [178]. Therefore, it is understandable that molecular cloning was chosen in order to increase the availability of interferon and to lower production costs.

Similar to insulin two different strategies were used:

1) Several working groups succeeded in isolating interferon-coding-DNA-segments via cDNA synthesis from mRNA of stimulated leukocytes and fibroblasts [168, 179].

2) Very recently, the total chemical synthesis of an interferon gene was reported. The nucleotide sequence was not deduced from the protein sequence of a natural interferon protein (which could not be determined), but was composed according to the sequencing data from the several interferon genes already known [173].

In this case, the synthetic gene was constructed to fulfil the following prerequisites:

1) The homologous initiation and termination signals and in addition restriction sites for integration into a vector were provided.

2) The actual gene sequence was changed in relation to the natural one to supply the enzymes of the host with the optimal codon composition because higher codon/anticodon pairing energies seem to greatly affect the level of expression. However, there is still one handicap with bacterially produced interferon.

Since *E. coli* as the host organism is not able to modify the protein post-translationally in the same way as it is modified in the mammalian cell (i.e. by glycosylation), it is not yet evident whether this difference will decrease the activity of the protein in vivo [180].

Urokinase is a mammalian enzyme which is essential for dissolving blood clots. It activates plasminogen to plasmin which then is able to degrade the fibrin skeleton of the clots. Urokinase is isolated either from human urine or from specific mammalian tissues in a rather elaborate procedure [181]. Therefore, this enzyme is not only available in insufficient amounts but also too expensive for a world-wide use. Streptokinase of bacterial origin, albeit easier and less expensive to obtain, is not a potent substitute because it is less specific and may show vaseolytic activity.

From the many efforts to obtain urokinase via molecular cloning quite recently one seems to have been successful, at least under laboratory conditions, in using a cDNA copy of the gene and testing enzymatic activity in a biological study [169].

Viral proteins originating mostly from the coat, act as antigenic determinants [182].

Thus, inactivated viruses are frequently used as vaccines for immunization. In order to avoid occasional accidents caused by viral contaminations, experiments are in progress to obtain coat proteins which may replace the viruses in vaccination via molecular cloning.

The major antigenes of hepatitis B virus and of foot and mouth disease virus were cloned via the cDNA technique and expressed in *E. coli* [170, 171].

However, as is always the case when cloned proteins are isolated from *E. coli*, a contamination with bacterial proteins, especially with enterotoxins, can hardly be avoided (see p. 156). This naturally may lead to complications when cloned proteins are used in human therapy.

Quite recently, it could be shown that bacterially produced antigenic protein of foot and mouth disease virus may act as vaccine in cattle and swine [183].

Somatotropin, also called human growth hormone, controls skeletal growth (of bones and cartilage). In therapy, treatment of hypopituitary dwarfism and other disorders such as bone fractures, skin burns and bleeding ulcers is required. Since somatotropin is specific to the species it is only available from pituitaries of human corpses. Recently, rather successful efforts were undertaken in using the cDNA technique to clone the growth hormone after insertion into the β-galactosidase gene of the lac operon of *E. coli* [167]. This bacterial protein has now been shown to be biologically active [184].

Somatostatin which is widely distributed throughout the brain and in places outside the central nervous system acts as inhibitor of somatotropin secretion and usually also inhibits a wide range of other hormones (such as thyrotropin, prolactin, ACTH, insulin, glucagon) [185]. As it has been recognized as therapeutical approach in various endocrine, gastrointestinal, and neuropsychiatric disorders, efforts to make it available via molecular cloning have been encouraged. Since somatostatin is a peptide consisting of 14 amino acids only, its genetic information could be synthesized chemically and cloned in *E. coli* after integration into the β-galactosidase gene [129].

This was the first case that a chemically synthesized gene was successfully cloned and subsequently expressed in *E. coli*.

β-*Endorphin* is a morphine-like polypeptide composed of 31 amino acids. It is produced in the pituitary gland and is therapeutically used for the treatment of chronical pains and as psychopharmacon [186]. The gene for β-endorphin was cloned as a cDNA copy from the mRNA transcript of the mouse [145].

Thymosin α_1 is one of the peptide hormones produced in the thymus gland and consists of 28 amino acids [187]. Its main function and perspective therapeutical benefit is the activation of T-cells and reconstitution of immune functions in humans suffering from certain immunodefiencies and also immunosuppressed conditions. Like somatostatin this hormone was cloned from a chemically synthesized gene in *E. coli* [172].

6.1.2 *Bacillus subtilis*

Among the gram-positive bacteria *B. subtilis*, the hay bacillus, is biochemically and genetically the most thoroughly analyzed organism [188]. There are several reasons for developing *B. subtilis* as a host in molecular cloning as an alternative to *E. coli*.

1) In fundamental research it is of interest to reveal differences in gene structure and gene expression between gram-negative and gram-positive bacteria [157, 189].

2) *B. subtilis* does not produce toxins. It is even used as human food in some countries (*B. subtilis* (natto) in Japan). Thus contamination of a prospective product has not to be regarded as in *E. coli* (p. 156) [189].

3) *B. subtilis* is already widely used in the industrial production of exoenzymes and antibiotics [190].

However, there are two main limits to the general use of *B. subtilis* in biotechnology which have not yet been completely overcome.

1) The coexistence of the entering DNA with the host DNA is very often impaired [191, 192].

2) Transformed *B. subtilis* strains are rather unstable in that they may loose the heterologous DNA [147].

However, cloning has been successful in some cases (see compilation of Table 5) mostly involving genes of related, i.e. gram-positive, bacteria such as *Staphylococcus* and infrequently heterologous prokaryotic and eukaryotic genes.

One example to be further discussed is the cloning and expression of cDNA copies of the antigenic determinants of two viruses, hepatitis-B-virus and foot-and-mouth disease virus [194]. Expression was achieved by a method analogous to that discussed for *E. coli* (see p. 159). i.e. inserting the foreign cDNA into a homologous gene (in this case, an erythromycin resistance gene) and thus using its transcription and translation signals. If such strains prove stable, they offer perspectives for industrial exploitation.

6.1.3 Other Bacteria

Besides *E. coli* and *B. subtilis* there are a few more bacteria used as host organisms for molecular cloning. Albeit these systems are not yet used for the cloning and expression of heterologous DNA, they have to be mentioned in this context as they probably have future perspectives.

One of the most interesting projects is that involving *soil bacteria* with the aim to transfer genes for nitrogen fixation to bacteria of the rhizosphere [196, 197].

Nitrogen is an essential factor for the growth of higher plants which are not able to assimilate nitrogen directly, but only in combined form, mostly as nitrate. The nitrogen compounds in the soil are subject to a continuous turnover so that nitrate is constantly formed from decaying

Table 5. Examples of heterologous molecular cloning in *Bacillus subtilis* and *Streptomyces lividans*

Host organism	Source of DNA	Gene or gene function	Ref.
Bacillus subtilis	*Escherichia coli*	Thymidylate synthetase	193)
	Hepatitis-B-virus	Core antigen	194)
	Foot-and-mouth disease virus	Coat protein VP1	
Streptomyces lividans	*E. coli*	Chloramphenicol resistance	195)
		Kanamycin resistance	

material or excretory end products. But agriculturally used soil has to be supplied additionally with nitrogen compounds. Principally, there are two ways: Firstly, by naturally fertilizing the soil with the help of nitrogen-fixing microorganisms, a method discussed already at the beginning of this century [198], and secondly by artificially fertilizing it with nitrate or industrially produced ammonia (Haber-Bosch process). However, since the latter process depends on fossil energy, it seems to become uneconomical in the long run and efforts are being made to find alternatives to artificial fertilizers, in other words, to find possibilities to achieve efficient nitrogen fixation by biological means [199].

On the one hand, there are plants (the *Leguminosae* syn. *Fabaceae*) living in symbiosis with nitrogen-fixing bacteria and on the other hand, there is a range of bacteria (and other microorganisms) in the rhizosphere of plants which are able to fix molecular nitrogen [200]. The leguminosae may be used as model system for plant-bacteria interactions (experiments making use of natural interactions will be discussed later). Another strategy is to transfer nitrogen-fixing abilities to other bacteria that live in close proximity to the roots of (crop) plants so that nitrogen compounds are available in sufficient amounts within the rhizosphere [201].

The first steps to reach this aim have already been made:

1) The genes responsible for nitrogen fixation were identified and isolated from *Klebsiella pneumoniae*, a soil-living, facultative anaerobe bacterium related to *E. coli* [202].

2) The genes were shown to be organized in a cluster (consisting of 7 nif-operons) and were transferred via a resistance plasmid to *E. coli* [203, 204].

3) As a prerequisite for the manipulation of its control the regulation of gene expression of the nif-operon was investigated [205, 206].

    Thus, mutants or in vitro manipulated nif-operons might be constructed that will show constitutive expression of the nif-genes.

4) The nif-genes were transferred to soil bacteria (*Enterobacter*) and few stable transformants could be isolated which had acquired the ability to fix nitrogen [196].

However, there remain many problems to be solved before the application of this strategy. For instance, it was found that besides the lack of stability of the nif$^+$-character in soil bacteria (*Enterobacteriaceae*) those retaining the genes (presumably integrated into the chromosome) do not excrete the bound nitrogen [207].

Furthermore, soil bacteria most prominent in number are *Arthrobacter* species [201], gram-positive (like *Bacillus subtilis*) and thus much more difficult to work with than species related to the gram-negative *E. coli*.

Among the bacteria also used for molecular cloning two more should be mentioned in this context:

1) *Pseudomonas* species (gram-negative) are characterized by their ability to synthesize a wide range of industrially important organic compounds. Furthermore, they may utilize a variety of heterocyclic or aromatic substrates that are not attacked by other bacteria. The genetic information concerned is mostly plasmid-encoded [208].

2) *Streptomyces* species (gram-positive) showing mycelial growth are industrially important as producers of various antibiotics [209].

For both bacteria vector molecules were constructed: for *Pseudomonas* based on plasmid DNA [210]; for *Streptomyces* based on both, plasmid and phage DNA [211-213]. The significance of these organisms in applied molecular cloning has yet to be established.

6.2 Eukaryotes

6.2.1 Fungi

The main aspect for searching an eukaryote for molecular cloning is the need for an organism that is able to deal with eukaryotic genes properly in contrast to *E. coli* that can neither splice mosaic genes nor recognize eukaryotic transcription and translation signals (as discussed above). Efforts to develop such a system have been most successful so far with the yeast *Saccharomyces cerevisiae* [49, 149]. There are several reasons for using yeasts as host organisms:

1) Yeasts are well analyzed genetically [214]. The knowledge of chromosomal genetics (tetrad analysis) and extrachromosomal genetics (mitochondrial inheritance) may be used to complement manipulations of in vitro recombination. Furthermore, a cytoplasmic plasmid DNA (2 μm DNA) is known and thoroughly studied in *S. cerevisiae* [47, 48, 215].

2) Yeasts as microorganisms are well suited for industrial purposes, since as unicellular organisms they are better adapted for growth in a bioreactor than multicellular organisms (e.g. some filamentous fungi).

3) Yeasts are non-toxic (see also p. 156) and used as food or food additive, e.g. in bread, beer, and wine production.

Heterologous gene expression in *S. cerevisiae*, as reported so far, is compiled in Table 6.

Cloning vectors used at present are either developed from the replicon of the 2 μm DNA plasmid (e.g. [49]) or from chromosomal origins of replication [63, 64] (see also p. 147).

Though the methods and techniques for cloning in yeast are rather established there are still some problems:

1) The transformed organisms usually show considerable instability of the introduced DNA [224]. This instability may be eliminated either by integrating the foreign gene into the chromosome [13] (this process probably results in a decreased level of

Table 6. Examples of heterologous molecular cloning in yeasts

Host organism	Source of DNA	Gene or gene function	Ref.
Saccharomyces cerevisiae			
direct cloning	*Escherichia coli*	Ampicillin resistance	84)
		Chloramphenicol resistance	84)
		Kanamycin resistance	84)
		β-Galactosidase	216)
		G418 resistance	217)
	Kluyveromyces lactis	β-Galactosidase	218)
	Drosophila melanogaster	Adenine metabolism	219)
indirect cloning	Human	Interferon	220)
	Chicken	Ovalbumin	221)
Saccharomycopsis lipolytica			
	S. cerevisia	his3, histidine metabolism	222)
Schizosaccharomyces pombe			
	S. cerevisiae	leu2, leucine metabolism	223)

expression) [225] or by using an artificial minichromosome (with a chromosomal centromere in addition to a functional replicon) which possesses mitotic and meiotic stability [160].

2) *S. cerevisiae* fails to express some of the eukaryotic genes introduced by direct cloning as shown for a rabbit mosaic gene [226]. This may be avoided by using the more complicated and time-consuming strategy of indirect cloning via cDNA [221].

From this point of view it seems almost natural that higher eukaryotes, which might be more suitable for direct expression of cloned foreign genes, are used as hosts for molecular cloning.

Organisms with potential perspectives are e.g. higher differentiated fungi like the hyphal fungi *Neurospora* and *Podospora*.

For *Neurospora* vectors containing a chromosomal gene as marker were constructed [227, 228]. Introduction of the vector into the cell is followed by chromosomal integration of the DNA due to the lack of an appropriate, i.e. homologous replicon.

In *Podospora* transformation using a vector derived from a mitochondrial plasmid-like plDNA was successful (see p. 147 [229]). Transformed colonies could be identified by premature aging included by plDNA. It could also be shown that the segment of mitochondrial DNA homologous to plDNA is capable of transformation, too (see p. 148) [65, 89].

There is up to now one example of heterologous gene expression in hyphal fungi, i.e. expression of the bacterial gene for β-lactamase in *Podospora* [65].

Molecular cloning has also been reported to occur in some yeasts other than *S. cerevisiae*, since it could be shown that replicons of *S. cerevisiae* are active (i.e. initiate replication) in *Saccharomycopsis lipolytica* [222] and *Schizosaccharomyces pombe* [223]. Some examples of heterologous gene expression in these hosts are listed in Table 6.

6.2.2 Plants

The overall aspect of molecular cloning in plants is to extend the natural gene pool of crop plants with the main aims of
— improving quality and quantity of yield
— breeding resistant lines
— developing nitrogen fixing plants.

There are several approaches to transform plants, one of which is mentioned in some detail in this context, i.e. molecular cloning with *Agrobacterium tumefaciens*. A review of molecular cloning in plants has recently been given by Cocking et al. [24].

A. tumefaciens is a gram-negative bacterium which induces crown gall tumours in a variety of dicotyledonous plants [230]. The molecular basis for tumor induction and bacterial growth in the plant cell is a plasmid of *A. tumefaciens*, the Ti plasmid (*tumor inducing plasmid*) [231] which, after transfer to the plant cell and integration of part of it into its genome [232], induces the synthesis of specific amino acids (the opines) which are essential for bacterial growth [233]. This phenomenon, which is a kind of parasitism where the parasite completely changes the genetic program of the host cell, has been called "genetic colonisation" [234].

The T-DNA (Tumor DNA) which is part of the Ti plasmid and integrated into the plant genome carries the determinants for opine synthesis, oncogenicity, and transfer [235, 236].

The strategy to use this natural system of heterologous cloning for molecular cloning in plants consists in co-transferring a selected DNA fragment (gene) with the T-DNA [237]. To achieve successful cloning it must then be ensured that the foreign DNA is replicated in a stable manner, expressed in the host cell, and inherited to daughter cells.

Up to now it could be shown that whole plants may be regenerated from tumorous cells which still contain parts of T-DNA (including the region of opine synthesis which serves as a marker for the presence of T-DNA) and which even retain this DNA in crosses (i.e. after sexual propagation) [238, 239]. The main efforts are now being made in manipulating T-DNA in such a way that it still contains the information for transfer and opine synthesis but loses the ability to induce oncogenicity, and may thus serve as a vector for foreign genes [238, 240].

This strategy utilizes the natural process of infection by *A. tumefaciens* involving transfer of the modified Ti plasmids to the plant cell [241]. It was already applied to transfer a bacterial transposon integrated into the T-DNA to the host plant cell [240, 242].

One more system of plant molecular cloning needs to be mentioned in this context, i.e. the plant viruses. Of these most attention is paid to the caulimoviruses (double-stranded DNA viruses, e.g. *cauli*flower *mos*aic virus) and the gemini viruses (single-stranded DNA viruses, e.g. bean golden mosaic virus) [243–245].

But problems of integration of foreign DNA and stable maintenance of the vector in the plant cell have not been thoroughly studied so far [246].

However, molecular cloning in plants suffers from three main handicaps [24]:
1) Cell lines which possess an easily selectable marker to distinguish transformed from non-transformed protoplasts are not readily available for each species.
2) Plant genes suitable and desirable for being cloned are largely missing at present.
3) Regeneration of whole plants from protoplasts is still restricted to a few genera excluding most crop plants of economic value (e.g. cereals).
4) Monocotyledonous plants are not susceptible to *Agrobacterium* infection and may thus not be manipulated by in vivo transformation.

6.2.3 Mammalian Cell Systems

The main aspect of molecular cloning in mammalian cell systems concerns fundamental research. The corresponding methods were mainly developed to investigate gene regulation and expression [247]. Heterologous molecular cloning in mammalian cells is not worth being studied especially with respect to applied research, because cultivation is too expensive [248] and yields are too low for industrial applications.

There are three techniques applied in general to transfer DNA to cells in culture:
1) Microinjection of purified DNA segments into isolated cells.

DNA is mostly injected into the nucleus, and stable transformation depends on subsequent recombination with the nuclear DNA [249–251].

2) Direct transformation of cells by DNA precipitated with calcium phosphate.

This method yields a very low rate of transformation. Moreover, it depends on recombination between transforming DNA and cellular DNA [252, 253].

3) Transfer of genetic information via viral genomes.

This mostly occurs with the help of the simian virus SV 40 which is a double-stranded DNA virus. SV 40 is capable of undergoing two different growth cycles, depending on the origin of its host cell. In permissive cells it follows the lytic cycle which results in rapid multiplication of the viral particles and lysis of the host cell. In non-permissive cells the viral genome is integrated into the cellular genome and replicated along with it [254, 255]. Both these properties of SV 40 were utilized for molecular cloning. Cultivation of the virus in permissive cells leads to a rapid propagation of the virus but also to lysis of the host cell [256, 257] whereas non-permissive cells only accommodate few copies of the viral genome which impedes expression of transferred genes [255, 258].

Summing up, it may be said that molecular cloning in mammalian cells is possible but impracticable for more general use because of lack of an autonomously replicating vector and of expensive cultivation.

Direct cloning and expression of heterologous genes as analyzed so far seem to be generally possible within the mammalian system, and expression of several prokaryotic genes has also been reported (e.g. [258–260]).

7 Conclusions and Perspectives

This chapter briefly summarizes the results of molecular cloning in the various systems and their potentials for industrial applications.

Table 7 gives a concise survey of the results of heterologous molecular cloning and gene expression compiled according to the host organism and the source of the foreign DNA. *Escherichia coli* appears to be a universal host for molecular cloning since expression of heterologous genetic information is almost generally possible either upon direct cloning (see Table 7) or upon indirect cloning (see p. 153). *Bacillus subtilis* in contrast does not very well express foreign genes which is partly due to the instability of the transformed strain, a problem that similarly arises with *S. cerevisiae*. However, for cloning of eukaryotic DNA *Saccharomyces cerevisiae* seems to be the ideal host up to now, albeit there are still some problems to be solved in direct cloning of mosaic genes (p. 164).

When discussing molecular cloning and its application some comments on economics have to be made. In a free economy a new technique has to prove competitive with respect to tested methods of production and it will only be advantageous for products which are expensive and available only to a limited extent at present. Especially, in cloning with *E. coli* high expenses will arise from extensive purification procedures of the protein product.

Another question is whether results obtained on a laboratory scale may be transferred to large scale industrial production, which is often hindered by a lack of stability of the strain (see p. 156).

Molecular cloning with plants may become relevant to agricultural application when it has been possible to develop improved strains which will retain the transferred character invariable and will breed viable progeny. Cell culture techniques finally will probably remain restricted to applications in fundamental research, due to expensive cultivation conditions.

Thus, molecular cloning is extremely valuable and rather successful under laboratory conditions, but it must be considered for each case individually whether this

Table 7. Summary of currently employed host organism and their potentials in molecular cloning
Explication of signs: + possible; (+) not generally possible; ± only exceptionally possible

Cloning system	Source of DNA	Stable replication		Trans-cription	Translation
		after chromosomal integration	as autonomous replicon		
Escherichia coli	Gram-negative bacteria		+	+	+
	Gram-positive bacteria		+	+	+
	Fungi		+	(+)	(+)
	Mammals		±	±	±
Bacillus subtilis	Gram-negative bacteria	+	(+)	±	±
Saccharomyces cerevisiae	Gram-negative bacteria	+	(+)	+	+
	Mammals	+	(+)	±	±
Mammalian cell	Gram-negative bacteria	+	(+)	(+)	(+)
	Heterologous mammals	+	(+)	+	+

technique is the method of choice for industrial applications. Furthermore, one has to be aware that it is just one genetic technique among the established possibilities of genetic manipulation.

One should not forget that "classical" recombination of DNA during the sexual or parasexual cycle is in many cases easier to perform, less elaborate and leads to more stable recombinants.

8 References

1. Esser, K., Kuenen, R.: Genetics of fungi, Berlin Heidelberg New York: Springer 1967
2. Whitehouse, H. L. K.: Towards an understanding of the mechanism of heredity, London: E. Arnold 1973
3. Stahl, F. W.: Genetic recombination, San Francisco: W. H. Freeman and Company 1979
4. Ball, C.: Genetic modification in filamentous fungi, in: Fungal Biotechnology. (Smith, J. E., Berry, D. R., Kristiansen, B. eds.), p. 43, The British Mycological Soc. Symp. Series No. 3, London: Academic Press 1980
5. Esser, K.: Impact of basic research on the practical application of fungal processes, in: Developments in Industrial Microbiology, Vol. 22, p. 19, Arlington (Virginia): Soc. for Industrial Microbiology 1981
6. Minuth, W., Esser, K.: unpublished data
7. Ephrussi, B., Weiss, M.: Proc. Nat. Acad. Sci. US 53, 1040 (1965)
8. Webber, H. J.: Science 18, 501 (1903)
9. Rieger, R., Michaelis, A., Green, M. M.: Glossary of genetics and cytogenetics, Berlin—Heidelberg—New York: Springer 1976
10. Cohen, S. N. et al.: Proc. Nat. Acad. Sci. US 70, 3240 (1973)
11. Ehrlich, S. D. et al.: Proc. Nat. Acad. Sci. 73, 4145 (1976)
12. Cohen, S. N. et al.: Macro- and microevolution of bacterial plasmids, in: Microbiology. (Schlesinger, D. ed.), p. 217, Washington, D.C.: Amer. Soc. of Microbiology 1978

13. Struhl, K. et al.: Proc. Nat. Acad. Sci. US 76, 1035 (1979)
14. Primrose, S. B., Ehrlich, S. D.: Plasmid 6, 193 (1981)
15. Collins, J.: Current Topics Microbiol. Immunol. 78, 121 (1976)
16. Curtiss, R.: Ann. Rev. Microbiol. 30, 507 (1976)
17. Klingmüller, W.: Genmanipulation und Gentherapie, Berlin Heidelberg New York: Springer 1976
18. Bukhari, A. I., Shapiro, J. A., Adhya, S. L.: DNA insertion elements, plasmids and episomes, Cold Spring Harbor Lab. 1977
19. Sinsheimer, R. L.: Ann. Rev. Biochem. 46, 415 (1977)
20. Colowick, S. P., Kaplan, N. D. (eds.): Methods in enzymology, Vol. 68, New York—London: Academic Press 1979
21. Klingmüller, W.: Naturwiss. 66, 182 (1979)
22. Hollenberg, C. P.: Progr. Botan. 42, 171 (1980)
23. Old, R. W., Primore, S. B.: Principles of gene manipulation. Studies in microbiology, Vol. 2, Oxford: Blackwell Scientific Publications 1980
24. Cocking, E. C. et al.: Nature 293, 265 (1981)
25. Esser, K., Stahl, U.: Hybridization, in: Biotechnology. (Rehm, H. J., Reed, G. eds.), p. 305, Weinheim: Verlag Chemie 1981
26. Griffith, F.: J. Hyg. 27, 113 (1928)
27. Avery, O. T., MacLeod, C. M., McCarthy, M.: J. Exptl. Med. 79, 137 (1944)
28. Spizizen, J.: Proc. Nat. Acad. Sci. US 44, 1072 (1958)
29. Ravin, A. W.: Adv. Genet. 10, 61 (1961)
30. Benoit, J. et al.: C.R. Acad. Sci. (Paris) 244, 2320 (1957)
31. Fox, A. S., Yoon, S. B.: Genetics 53, 897 (1966)
32. Hess, D.: Z. Pflanzenphysiol. 60, 348 (1969)
33. Benoit, J. et al.: Trans. N.Y. Acad. Sci. 22, 494 (1960)
34. Mishra, N. C., Tatum, E. L.: Proc. Nat. Acad. Sci. US 70, 3875 (1973)
35. Mishra, N. C., Szabo, G., Tatum, E. L.: Nucleic-acid induced genetic changes in Neurospora, in: The role of RNA in Reproduction and Development. (Niu, M. C., and Segal, S. J. eds.), North-Holland Publ. Co. 1973
36. Arber, W., Dussoix, D.: J. Molec. Biol. 5, 18 (1962)
37. Dussoix, D., Arber, W.: J. Molec. Biol. 5, 37 (1962)
38. Jackson, D. A., Symonds, R. H., Berg, P.: Proc. Nat. Acad. Sci. US 69, 2904 (1972)
39. Southern, E. M.: J. Molec. Biol. 98, 503 (1975)
40. Esser, K., Blaich, R.: Adv. Genet. 17, 107 (1973)
41. Dubnau, D. et al.: Molecular cloning in Bacillus subtilis, in: Genetic Engineering, Vol. 2 (Setlow, J. K., Hollaender, A. eds.), p. 115, New York—London: Plenum Press 1980
42. Novick, R. P. et al.: Bacteriol. Rev. 40, 168 (1976)
43. Jacob, F., Brenner, S., Cuzin, F.: Cold Spring Harbor Symp. Quant. Biol. 28, 329 (1963)
44. Lederberg, J.: Physiol. Rev. 32, 403 (1952)
45. Hershey, A. D. (ed.): The bacteriophaga lambda, New York: Cold Spring Harbor Laboratory 1971
46. Das, G. C., Niyogi, S. K.: Progr. Nucl. Acid Res. Molec. Biol. 25, 187 (1981)
47. Hartley, J. L., Donelson, J. E.: Nature 286, 860 (1980)
48. Hollenberg, C. P.. Borst, P., Van Bruggen, E. F. J.: Biochim. Biophys. Acta 209, 1 (1970)
49. Beggs, J.: Nature 275, 104 (1978)
50. Livingston, D. M., Kupfer, D. M.: J. Molec. Biol. 116, 249 (1977)
51. Taketo, M., Jazwinski, S. M., Edelman, G. M.: Proc. Nat. Acad. Sci. US 77, 3144 (1980)
52. Stahl, U. et al.: Molec. gen. Genet. 162, 341 (1978)
53. Sinclair, J. H. et al.: Science 156, 1234 (1967)
54. Del Giudice, L. et al.: Molec. gen. Genet. 172, 165 (1979)
55. Del Giudice, L. et al.: Molec. gen. Genet. 181, 306 (1981)
56. Locker, J., Lewin, A., Rabinowitz, M.: Plasmid 2, 155 (1979)
57. Mannella, C. A., Goewert, R. R., Lambowitz, A. M.: Cell 18, 1197 (1979)
58. De Vries, H. et al.: Current Genet. 3, 205 (1981)
59. Lazarus, C. M. et al.: Eur. J. Biochem. 106, 633 (1980)

60. Pring, D. R. et al.: Proc. Nat. Acad. Sci. US *74*, 2904 (1977)
61. Kemble, R. J., Bedbrooke, J. R.: Nature *284*, 565 (1980)
62. Collins, R. A. et al.: Cell *24*, 443 (1981)
63. Stinchcomb, D. T., Struhl, K., Davis, R. W.: Nature *282*, 39 (1979)
64. Beach, D., Piper, M., Shall, S.: Nature *284*, 185 (1980)
65. Stahl, U. et al.: Proc. Nat. Acad. Sci. US *79*, 3641 (1982)
66. Tiollais, P., Rambach, A.: La Recherche *8*, 821 (1977)
67. Cohen, S. N. et al.: Proc. Nat. Acad. Sci. US *70*, 3240 (1973)
68. Hershfield, V. et al.: Proc. Nat. Acad. Sci. US *71*, 3455 (1974)
69. Rodriguez, R. L. et al.: Construction and characterization of cloning vehicles, in: ICN-UCLA Symp. Molec. Cell Biol., Vol. V, (Nierlich, D. P., Rutter, W. J., Fox, C. F. eds.), New York: Academic Press 1976
70. Bolivar, F. et al.: Gene *2*, 75 (1977)
71. Bolivar, F. et al.: Gene *2*, 95 (1977)
72. Bolivar, F.: Gene *4*, 121 (1978)
73. Chang, A. C. Y., Cohen, S. N.: J. Bacteriol. *134*, 1141 (1978)
74. Tiemeier, D., Enquist, L., Leder, P.: Nature *263*, 526 (1976)
75. Messing, J. et al.: Nucl. Acids Res. *9*, 309 (1981)
76. Hohn, B., Collins, J.: Gene *11*, 291 (1980)
77. Bernhard, K., Schrempf, H., Goebel, W.: J. Bacteriol. *133*, 897 (1978)
78. Ehrlich, S. D.: Proc. Nat. Acad. Sci. US *74*, 1680 (1977)
79. Gryczan, T. J., Contente, S., Dubnau, D.: J. Bacteriol. *134*, 318 (1978)
80. Weisblum, B. et al.: J. Bacteriol. *137*, 635 (1979)
81. Kreft, J., Bernhard, K., Goebel, W.: Molec. gen. Genet. *162*, 59 (1978)
82. Chang, S., Cohen, S. N.: Molec. gen. Genet. *168*, 111 (1979)
83. Duncan, C. H., Wilson, G. A., Young, F. E.: Proc. Nat. Acad. Sci. US *75*, 3664 (1978)
84. Hollenberg, C. P.: The expression of bacterial resistance genes in the yeast *Saccharomyces cerevisiae*, in: Plasmids of Medical, Environmental and Commercial Importance. (Timmis, K. N., Pühler, A. eds.), p. 481, Amsterdam—New York—Oxford: Elsevier/North-Holland Medical Press 1979
85. Kingsman, A. J. et al.: Gene *7*, 141 (1979)
86. Broach, J. R., Strathern, J. N., Hicks, J. B.: Gene *8*, 121 (1979)
87. McKnight, S. L., Croce, C., Kingsbury, R.: Carnegie Institute Year Book *78*, 56 (1979)
88. Stahl, U. et al.: Molec. gen. Genet. *178*, 639 (1980)
89. Kück, U., Stahl, U., Esser, K.: Current Genet. *3*, 151 (1981)
90. Murray, N. E., Murray, K.: Nature *251*, 476 (1974)
91. Williams, B. G., Blattner, F. R.: Bacteriophage lambda vectors for DNA cloning, in: Genetic Engineering, Principles and Methods (Setlow, J. K., Hollaender, A. eds.), p. 201, New York— London: Plenum Press 1980
92. Hohn, B.: J. Molec. Biol. *98*, 93 (1975)
93. Collins, J., Hohn, B.: Proc. Nat. Acad. Sci. US *75*, 4242 (1978)
94. Messing, J. et al.: Proc. Nat. Acad. Sci. US *74*, 3642 (1977)
95. Air, G. M.: Crit. Rev. Biochem. *6*, 1 (1979)
96. Watson, J.: Molecular biology of the gene, Menlo Park, Benjamin Inc. 1975
97. Breathnach, R., Mandel, J. L., Chambon, P.: Nature *270*, 314 (1977)
98. Gilbert, W.: Nature *271*, 501 (1978)
99. Gallwitz, D., Sures, I.: Proc. Nat. Acad. Sci. US *77*, 2546 (1980)
100. Durica, D. S., Schloss, J. A., Crain, W. R. jr.: Proc. Nat. Acad. Sci. US *77*, 5683 (1980)
101. Fyrberg, E. A., Kindle, K. L., Davidson, N.: Cell *19*, 365 (1980)
102. Goldberg, D. A.: Proc. Nat. Acad. Sci. US *77*, 5794 (1980)
103. Young, R. A., Hagenbüchle, O., Schibler, U.: Cell *23*, 451 (1981)
104. Slonimski, P. P. et al.: Mosaic organization and expression of the mitochondrical DNA region controlling cytochrome c reductase and oxidase. III. A model of structure and function, in: Biochemistry and Genetics of Yeast: Pure and Applied Aspects. (Bacila, M., Horecker, L., Stoppani, A. O. M. eds.), p. 391, New York: Academic Press 1978
105. Nunberg, J. H. et al.: Cell *19*, 355 (1980)
106. Tsujimoto, Y., Suzuki, Y.: Cell *16*, 425 (1979)

107. Jeffreys, A. J., Flavell, R. A.: Cell *12*, 1097 (1977)
108. Tilghman, S. M. et al.: Proc. Nat. Acad. Sci. US *75*, 725 (1978)
109. Mears, J. G. et al.: Cell *15*, 15 (1978)
110. Tonegawa, S. et al.: Proc. Nat. Acad. Sci. US *75*, 1485 (1978)
111. Lomedico, P. et al.: Cell *18*, 545 (1979)
112. Bell, G. I. et al.: Nature *284*, 26 (1980)
113. Sullivan, D. et al.: Nature *289*, 516 (1981)
114. Nguyen-Hun, M. C. et al.: Proc. Nat. Acad. Sci. US *76*, 76 (1979)
115. Catterall, J. F. et al.: Nature *278*, 323 (1979)
116. Sun, S. M., Slightom, J. L., T. C.: Nature *289*, 37 (1981)
117. Ryffel, G. U. et al.: Cell *19*, 53 (1980)
118. Nevins, R. J., Darnell, J. E.: Cell *15*, 1477 (1978)
119. Efstratiadis, A. et al.: Cell *7*, 279 (1976)
120. Higuchi, R. et al.: Proc. Nat. Acad. Sci. US *73*, 3146 (1976)
121. Goodman, H. M., MacDonald, R. J.: Cloning of hormone genes from a mixture of cDNA
     molecules, in: Methods in Enzymology. (Wu, R. ed.), Vol. 68, p. 75, New York: Academic Press
     1979
122. Cosloy, S. D., Oishi, M.: Proc. Nat. Acad. Sci. US *70*, 84 (1973)
123. Pribnow, D.: Proc. Nat. Acad. Sci. US *72*, 784 (1975)
124. Schaller, H., Gray, C., Herrmann, K.: Proc. Nat. Acad. Sci. US *72*, 737 (1975)
125. Gannon, F. et al.: Nature *278*, 428 (1979)
126. Flavell, R. A.: Nature *285*, 356 (1980)
127. Grosschedle, R., Birnstiel, M. L.: Proc. Nat. Acad. Sci. US *77*, 1432 (1980)
128. Benoist, C., Chambon, P.: Nature *290*, 304 (1981)
129. Itakura, K. et al.: Science *198*, 1056 (1977)
130. Shine, J., Dalgarno, L.: Proc. Nat. Acad. Sci. US *71*, 1342 (1974)
131. Rottman, F., Shatkin, A. J., Perry, R. P.: Cell *3*, 197 (1974)
132. Shatkin, A. J.: Cell *9*, 645 (1976)
133. Sippel, A. E. et al.: Proc. Nat. Acad. Sci. US *71*, 4635 (1974)
134. Marbaix, G. et al.: Proc. Nat. Acad. Sci. US *72*, 3065 (1975)
135. Hautala, J. A. et al.: Proc. Nat. Acad. Sci. US *76*, 5774 (1979)
136. Villa-Komaroff, L. et al.: Proc. Nat. Acad. Sci. US *75*, 3727 (1978)
137. Backman, K., Ptashne, M.: Cell *13*, 65 (1978)
138. Chang, A. C. Y. et al.: Nature *275*, 617 (1978)
139. Goeddel, D. V. et al.: Proc. Nat. Acad. Sci. US *76*, 106 (1979)
140. Itakura, K., Riggs, A. D.: Science *209*, 1401 (1980)
141. Köster, H.: Drug Research *30*, 548 (1980)
142. Barrell, B. G., Bankier, A. T., Drouin, J.: Nature *282*, 189 (1979)
143. Struhl, K., Davis, R. W.: Proc. Nat. Acad. Sci. US *74*, 5255 (1977)
144. Skalka, A., Shapiro, L.: Gene *1*, 65 (1976)
145. Shine, J. et al.: Nature *285*, 456 (1980)
146. Hitzeman, R. A., Clarke, L., Carbon, J.: J. Biol. Chem. *255*, 12073 (1980)
147. Goebel, W. et al.: Replication and gene expression of extrachromosomal replicons in unnatural
     bacterial hosts, in: Genetic Engineering. (Boyer, H. W., Nicosia, S. eds.), p. 47, Amsterdam—
     New York—Oxford: Elsevier Biomedical Press 1978
148. Dobson, M. J., Futcher, A. B., Cox, B. S.: Current Genet. *2*, 201 (1980)
149. Hinnen, A., Hicks, J. B., Fink, G. R.: Proc. Nat. Acad. Sci. US *75*, 1929 (1978)
150. Noack, D. et al.: Molec. gen. Genet. *184*, 121 (1981)
151. Sena, E. P. et al.: J. Bacteriol. *123*, 497 (1975)
152. Sherratt, D. J.: Cell *3*, 189 (1974)
153. Makler, I., Halvorson, H. O.: J. Bacteriol. *131*, 374 (1977)
154. Chi, N. W., Ehrlich, S. D., Lederberg, J.: J. Bacteriol. *133*, 816 (1978)
155. Brammar, W. J., Muir, S., McMorris, A.: Molec. gen. Genet. *178*, 217 (1980)
156. Ehrlich, S. D.: Proc. Nat. Acad. Sci. US *75*, 1433 (1978)
157. Ehrlich, S. D., Sgaramella, V.: Trends Biochem. Sci. *3*, 259 (1978)
158. Ratzkin, B., Carbon, J.: Proc. Nat. Acad. Sci. US *74*, 487 (1977)
159. Carbon, J. et al.: Brookhaven Symp. Biol. *29*, 277 (1977)

160. Clarke, L., Carbon, J.: Nature *287*, 504 (1980)
161. Hsiao, C., Carbon, J.: Proc. Nat. Acad. Sci. US *76*, 3829 (1979)
162. Citron, B. A., Feiss, M., Donelson, J. E.: Gene *6*, 251 (1979)
163. Crabéel, M. et al.: Arch. Int. Physiol. Biochim. *88*, B21 (1980)
164. Guerry-Kopecko, P., Wickner, R. B.: J. Bacteriol. *143*, 1530 (1980)
165. Vapnek, D. et al.: Proc. Nat. Acad. Sci. US *74*, 3508 (1977)
166. Kit, S. et al.: Proc. Nat. Acad. Sci. US *78*, 582 (1981)
167. Goeddel, D. V. et al.: Nature *281*, 544 (1979)
168. Nagata, S. et al.: Nature *284*, 316 (1980)
169. Ratzkin, B. et al.: Proc. Nat. Acad. Sci. US *78*, 3313 (1981)
170. Küpper, H. et al.: Nature *289*, 555 (1981)
171. Burrell, C. J. et al.: Nature *279*, 43 (1979)
172. Wetzel, R. et al.: Biochemistry *19*, 6069 (1980)
173. Edge, M.D. et al.: Nature *292*, 756 (1981)
174. Guarente, L. et al.: Cell *20*, 543 (1980)
175. Stewart, W. E. II: The interferon system, Berlin Heidelberg New York: Springer 1979
176. Newmark, P.: Nature *291*, 105 (1981)
177. Sun, M.: Science *212*, 141 (1981)
178. Cantell, K., Hirvonen, S.: Tex. Rep. Biol. Med. *35*, 138 (1977)
179. Derynck, R. et al.: Nature *287*, 193 (1980)
180. Stewart, W. E. II et al.: Gene *11*, 181 (1980)
181. Barlow, E. H. et al.: Production of plasminogen activator by tissue culture techniques, in: Proteases and Biological Control. (Reich, E., Rifkin, D. B., Shaw, E. eds.), p. 325, New York: Cold Spring Harbor Laboratories 1975
182. Langbeheim, H., Arnon, R., Sela, M.: Proc. Nat. Acad. Sci. US *73*, 4636 (1976)
183. Kleid, D. G. et al.: Science *214*, 1125 (1981)
184. Olson, K. C. et al.: Nature *293*, 408 (1981)
185. Vale, W., Rivier, C., Brown, M.: Ann. Rev. Physiol. *39*, 473 (1977)
186. Foley, K. M. et al.: Proc. Nat. Acad. Sci. US *76*, 5377 (1979)
187. Goldstein, A. L. et al.: Proc. Nat. Acad. Sci. US *74*, 725 (1977)
188. Young, F. E., Wilson, G. A.: Genetics of *Bacillus subtilis* and other gram-positive sporulating bacilli, in: Spores V. (Halvorson, H. O., Hanson, R., Campbell, L. L. eds.), p. 77, Washington, D.C.: Amer. Soc. for Microbiology 1972
189. Ehrlich, S. D. et al.: *Bacillus subtilis* as a host for DNA cloning, in: Genetic Engineering. (Boyer, H., Nicosia, S. eds.), p. 25, Amsterdam—New York—Oxford: Elsevier Biomedical Press 1978
190. Priest, F. G.: Bacteriol. Rev. *41*, 711 (1977)
191. De Vos, D. M. et al.: Molec. gen. Genet. *181*, 424 (1981)
192. Canosi, U., Iglesias, A., Trautner, T.: Molec. gen. Genet. *181*, 434 (1981)
193. Rubin, E. M., Wilson, G. A., Young, F. E.: Gene *10*, 227 (1980)
194. Hardy, K., Stahl, S., Küpper, H.: Nature *293*, 481 (1981)
195. Schottel, J. L., Bibb, M. J., Cohen, S. N.: J. Bacteriol. *146*, 360 (1981)
196. Kleeberger, A., Klingmüller, W.: Molec. gen. Genet. *180*, 621 (1980)
197. Klingmüller, W., Kleeberger, A.: Ber. Deutsch. Bot. Ges. Bd. *94*, 85 (1981)
198. Hiltner, L.: Brache, Arb. Deut. Landw. Ges. *98*, 59 (1904)
199. Anderson, K. et al.: Trends Biochem. Sci. *5*, 35 (1980)
200. Burns, R. C., Hardy, R. W.: Nitrogen fixation in bacteria and higher plants, Berlin—Heidelberg—New York: Springer 1975
201. Klingmüller, W.: Naturwiss. *66*, 182 (1979)
202. Streicher, S., Gurney, E., Valentine, R. V.: Proc. Nat. Acad. Sci. US *68*, 1174 (1971)
203. Cannon, F. C. et al.: Proc. Nat. Acad. Sci. US *74*, 2963 (1977)
204. Pühler, A., Burkhard, H. J., Klipp, W.: Molec. gen. Genet. *176*, 17 (1979)
205. Riedel, G. et al.: The nitrogen fixation (nif) operon of *Klebsiella pneumoniae*: Cloning nif-genes and the isolation of nif control mutants, in: Miami Winter Symp. Vol. 13 (Scott, W. A., Werner, R. eds.), p. 115, New York—San Francisco—London: Academic Press 1977
206. Buchanan-Wollaston, V. et al.: Nature *294*, 776 (1981)
207. Klingmüller, W.: personal communication

208. Chakrabarty, A. M.: Ann. Rev. Genet. *10*, 7 (1976)
209. Berdy, J.: Adv. Appl. Microbiol. *18*, 309 (1974)
210. Bagdasarian, M. et al.: Gene *16*, 237 (1981)
211. Bibb, M., Schottel, J. L., Cohen, S. N.: Nature *284*, 526 (1980)
212. Thompson, J. C., Ward, J. M., Hopwood, D. A: Nature *286*, 525 (1980)
213. Suarez, J. E., Chater, K. F.: Nature *286*, 527 (1980)
214. Petes, T. D.: Ann. Rev. Biochem. *49*, 845 (1980)
215. Guérineau, M.: Plasmid DNA in yeast; in: Viruses and Plasmids in Fungi. (Lemke, P. A. ed.),
     p. 540, New York—Basel: Marcel Dekker 1979
216. Panthier, J. et al.: Current Genet. *2*, 109 (1980)
217. Jimenez, A., Davies, J.: Nature *287*, 869 (1980)
218. Dickson, R. C.: Gene *10*, 347 (1980)
219. Henikoff, S. et al.: Nature *289*, 33 (1981)
220. Hitzeman, R. A. et al.: Nature *293*, 717 (1981)
221. Mercereau-Puijalon, O. et al.: Gene *11*, 163 (1980)
222. Ferreira, N. P., Mitchell, D. J., Thomson, J. A.: Transformation of *Saccharomycopsis lipolytica*
     by chimaeric plasmid DNA, in: Symp. on Yeast Genetics and Molecular Biology, p. 30,
     Louvain: 1980
223. Beach, D., Nurse, P.: Nature *290*, 140 (1981)
224. Gerbaud, C. et al.: Gene *5*, 233 (1979)
225. McNeil, J. B., Storms, R. K., Friesen, J. D.: Current Genet. *2*, 17 (1980)
226. Beggs, J. D. et al.: Nature *283*, 835 (1980)
227. Case, M. E. et al.: Proc. Nat. Acad. Sci. US *76*, 5259 (1979)
228. Schweizer, M. et al.: Gene *14*, 23 (1981)
229. Tudzynski, P., Stahl, U., Esser, K.: Current Genet. *2*, 181 (1980)
230. De Cleene, M., De Ley, J.: Bot. Rev. *42*, 389 (1976)
231. Watson, B. et al.: J. Bacteriol. *123*, 255 (1975)
232. Chilton, M. D. et al.: Cell *11*, 263 (1977)
233. Bomhoff, G. et al.: Molec. gen. Genet. *145*, 177 (1976)
234. Schell, J. et al.: Proc. Roy. Soc. (London) Ser. B *204*, 251 (1979)
235. Depicker, A., Van Montagu, M., Schell, J.: Nature *275*, 150 (1978)
236. Schell, J., Van Montagu, M.: New insight into the molecular genetics of fungi and plants, in:
     Recombinant DNA and Genetic Experimentation. (Morgan, J., Wheelan, W. J. eds.), p. 91,
     Oxford—New York: Pergamon Press 1979
237. Rörsch, A., Schilperoort, R. A.: *Agrobacterium tumefaciens* plasmids: potential vectors for gene-
     tic engineering of plants, in: Genetic Engineering. (Boyer, H. W., Nicosia, S. eds.), p. 189,
     Amsterdam—New York—Oxford: Elsevier Biomedical Press 1978
238. Otten, L. et al.: Molec. gen. Genet. *183*, 209 (1981)
239. Wullems, G. J. et al.: Proc. Nat. Acad. Sci. US *78*, 4344 (1981)
240. Ooms, G. et al.: Gene *14*, 33 (1981)
241. Marton, L. et al.: Nature *277*, 129 (1979)
242. Hernalsteens, J. et al.: Nature *287*, 654 (1980)
243. Shepherd, R. J.: Ann. Rev. Plant Physiol. *30*, 405 (1979)
244. Haber, S.: Nature *289*, 324 (1981)
245. Howell, S. H., Walker, L. L., Walden, R. M.: Nature *293*, 483 (1981)
246. Gronenborn, B. et al.: Nature *294*, 773 (1981)
247. Flavell, R. A.: Trends Biochem. Sci. *5*, 313 (1980)
248. Spier, R. E.: Adv. Chem. Eng. *14*, 119 (1980)
249. Gordon, J. W. et al.: Proc. Nat. Acad. Sci. US *77*, 7380 (1980)
250. Capechi, M. R.: Cell *22*, 479 (1980)
251. Anderson, W. F. et al.: Proc. Nat. Acad. Sci. US *77*, 5399 (1980)
252. Maitland, N. J., McDougall, J. K.: Cell *11*, 233 (1977)
253. Wigler, M. et al.: Cell *16*, 777 (1979)
254. Fiers, W. et al.: Nature *273*, 113 (1978)
255. Hamer, D. H.: DNA cloning in mammalian cells with SV40 vectors, in: Genetic Engineering,
     Principles and Methods, Vol. 2. (Setlow, J. K., Hollaender, A. eds.), p. 83, New York—London:
     Plenum Press 1980

256. Hamer, D. H. et al.: J. Molec. Biol. *112*, 155 (1977)
257. Mulligan, R. C., Howard, B. H., Berg, P.: Nature *277*, 108 (1979)
258. Mulligan, R. C., Berg, P.: Science *209*, 1422 (1980)
259. Wong, T., Nicolau, C., Hofschneider, P. H.: Gene *10*, 87 (1980)
260. O'Hare, K., Benoist, C., Breathnach, R.: Proc. Nat. Acad. Sci. US *78*, 1527 (1981)

Production of L-Tryptophan by Microbial Processes

László Nyeste, Miklós Pécs, Béla Sevella, János Holló
University of Technical Sciences, Budapest
Institute of Agricultural Chemical Technology, 1521 Budapest, Gellért tér 4. Hungary

Recently L-Tryptophan has acquired increasing significance in various fields of the chemical, animal feed and pharmaceutical industries.

For the production of L-tryptophan many kinds of synthetic, enzymatic and fermentation methods are known.

This review places emphasis on microbiological production methods though a brief description of other methods can also be found. After a brief discussion of L-tryptophan biosynthesis and its regulation, the various production methods, the significance of the substituted tryptophans, the main areas of its use and the situation in the tryptophan market will be discussed.

The main purpose of the authors is to review the knowledge of the past twenty years and to summarize this for researchers and industrial fermentation experts. The authors wish to give a widely useful outline of the field, including the less known East European and Soviet results as well. In the authors' opinion, the main question of industrial production of L-tryptophan in the near future will be the economy of production rather than bioengineering research itself.

1 Introduction

Several surveys have been published on the preparation of amino acids [1,2,3,4,5], but some of these became obsolate while others do not take East European and USSR publications into consideration. Hence, the publication of a paper dealing only with tryptophan, summarizing recent results in this field both for researchers and industrial experts, seems justified. Though this paper deals primarily with bioprocesses, it also offers a brief survey of other processes.

2 Biosynthesis of Tryptophan

Not all living organisms are able to biosynthesize L-Trp. Prototrophic microorganisms and higher plants are able to do so, but mammals, for which this amino acid is essential, are unable or only partly able to synthesize it. The steps of synthesis are identical for all living things, but the genes of tryptophan biosynthesis are organized differently in various species of organisms. Similarly, the enzymes catalyzing a given reaction step may be different in their character, aggregation and regulation.

The biochemical elaboration of the structure of compounds containing aromatic rings is a complicated, energy-requiring process. The synthesis of the biochemically needed molecules with aromatic ring systems proceeds according to a few thermo-dynamically possible schemes (formation of amino acids, purine ring, pyrimidine ring). The biosynthesis of aromatic amino acids also involves the production of the intermediates of several other compounds (e.g. p-amino-benzoate, p-hydroxy-benzoate, and even nicotinic acid and picolinic acid).

2.1 Common Aromatic Synthesis Pathway

Synthesis begins with the coupling of phosphoenol pyruvate (PEP) and D-erythrose-4-phosphate (E4P), two metabolites of high energy content in carbohydrate metabolism [Fig. 1, Step 1]. The product is a sugar derivative containing seven carbon atoms: 3-desoxy-D-arabino-heptulosonate-7-P (DAHP).

The next steps of the process were elucidated by mutational and isotope-labelled techniques [6,7,8].

DAHP is cyclizised by the action of the enzyme 5-dehydro-quinate-synthase [Step 2] and the aliphatic six membered ring of 5-dehydro-quinate (DHQ) is formed. The actual substrate for the enzyme is DAHP-1,7-diphosphate, which is formed as an intermediate of short life directly on the enzyme. DHQ is converted by loss of water into 5-dehydro-shikimate [Step 3]. The reaction proceeds easily, without energy input, because the three resulting conjugated double bonds form a mesomeric system, which is a state of low energy. The process, proceeding also spontaneously, is accelerated by the enzyme 5-dehydro-quinase.

The next step is energy-intensive and, for hydrogenation into shikimic acid (SA), the enzyme dehydro-shikimate-reductase needs $NADPH + H^+$ cosubstrate [Step 4]. The phosphorylation of shikimic acid [Step 5] and its subsequent reaction with phospho-enol-pyruvate [Step 6] yield 3-enol-pyruvyl-shikimic acid-5-P (EPSAP).

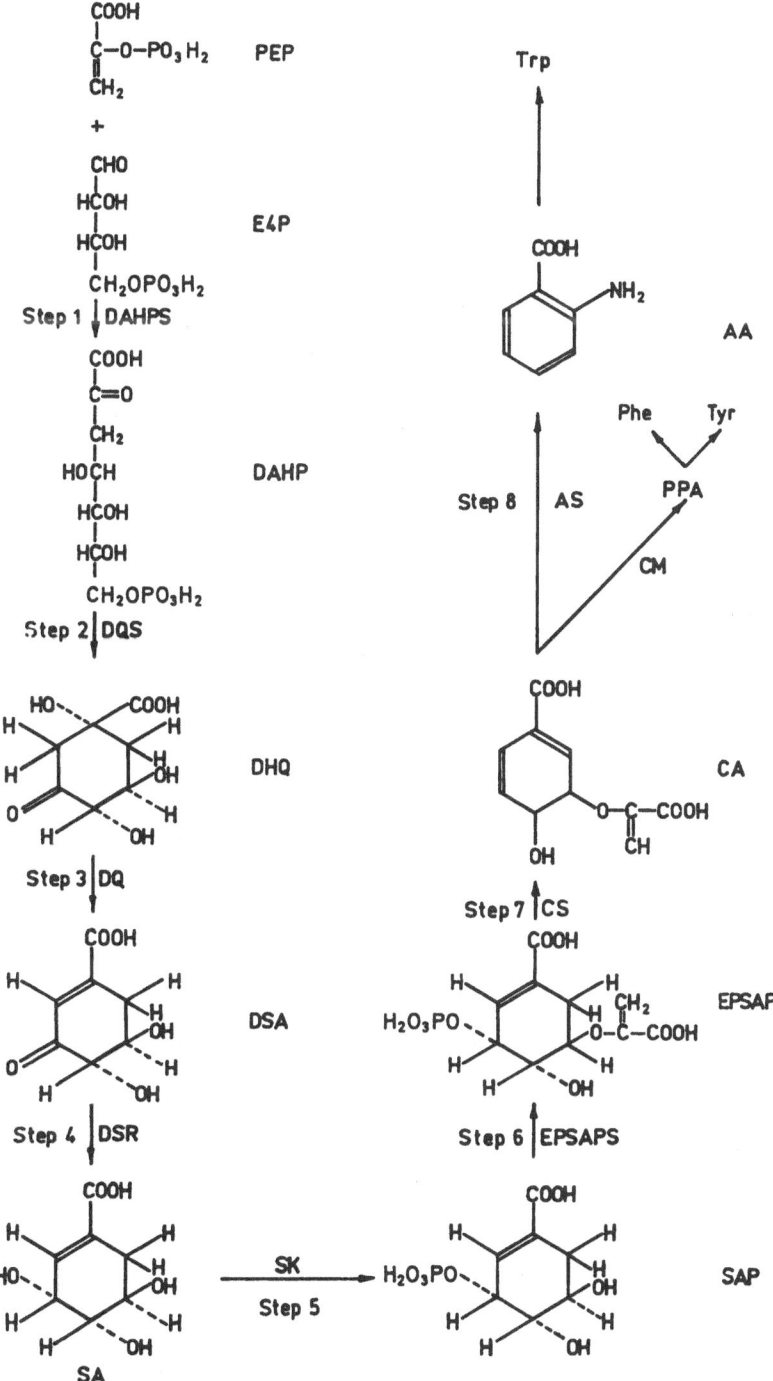

Fig. 1. Common biosynthetic pathway of aromatic amino acids

From this chorismic acid (CA) containing two double bonds is formed by the cleavage of phosphoric acid and rearrangement [Step 7]. In the more exact mechanism of the reaction a dehydrogenation process is presumably also involved, because FAD and an FAD-regenerating system are needed for the conversion. Chorismic acid is a very important intermediate in the biosynthesis of aromatic amino acids and the aromatic factors mentioned, because this is the branching point in the metabolic pathway. The molecule may enter reactions of several kinds, which are catalysed by different enzymes. In the direction of Phe and Tyr the synthesis is realized via prephenic acid as the common intermediate. In the direction of the biosynthesis of tryptophan the first step is the conversion of chorismic acid into anthranilic acid.

2.2 Biosynthesis of the Indole Skeleton

From the branching point, i.e. from chorismic acid, five further reaction steps lead to tryptophan, which can be considered as the end-product.

Anthranilate-synthase (AS) forms the aromatic structure and attaches an amino group to the ring [Step 8]. During the reaction pyruvic acid is set free. The donor of the amino group is glutamine under in vivo conditions, but in vitro it can be replaced by the ammonium ion.

The next conversion step [Step 9], condensation catalysed by phosphoribosyl-transferase (PRT), is an energy-extracting process. Ribose phosphate, attached to the aromatic amino group, participates in the reaction as ribose-5-phosphate-1-pyro-phosphate, and for the formation of this substrate, containing a bond of high energy, ATP is needed. The N-(5'-phosphoribosyl)-anthranilic acid (PRA) formed is converted by ring opening and by Amadori-rearrangement into 1-(o-carboxyphenyl-amino)-1-deoxyribulose-5-P [CDRP, Step 10].

This is followed by the formation of the hetero-aromatic second ring of the indole skeleton (In). The carbon atom, carrying the hydroxy group of enolic character of CDRP, is attached in ortho position to the benzene ring. The product formed is indole-3-glycerol phosphate (InGP), the structure of which is already very similar to that of tryptophan: a three-membered side-chain is attached to the indole skeleton [Step 11]. However, the terminating steps do not proceed by the conversion of the functional groups of the side-chain, but by the exchange of the complete side-chain [9].

2.3 The Tryptophan-synthase Reaction

Tryptophan-synthase (TS) catalyses two reactions:

$$TS(A) \quad InGP \quad \rightarrow In + glyceraldehyde\text{-}3\text{-}P$$

$$TS(B) \quad In + L\text{-}Ser \rightarrow L\text{-}Trp + H_2O$$

Thus, the overall reaction scheme is:

$$InGP + L\text{-}Ser \rightarrow L\text{-}Trp + glyceraldehyde\text{-}3\text{-}P + H_2O$$

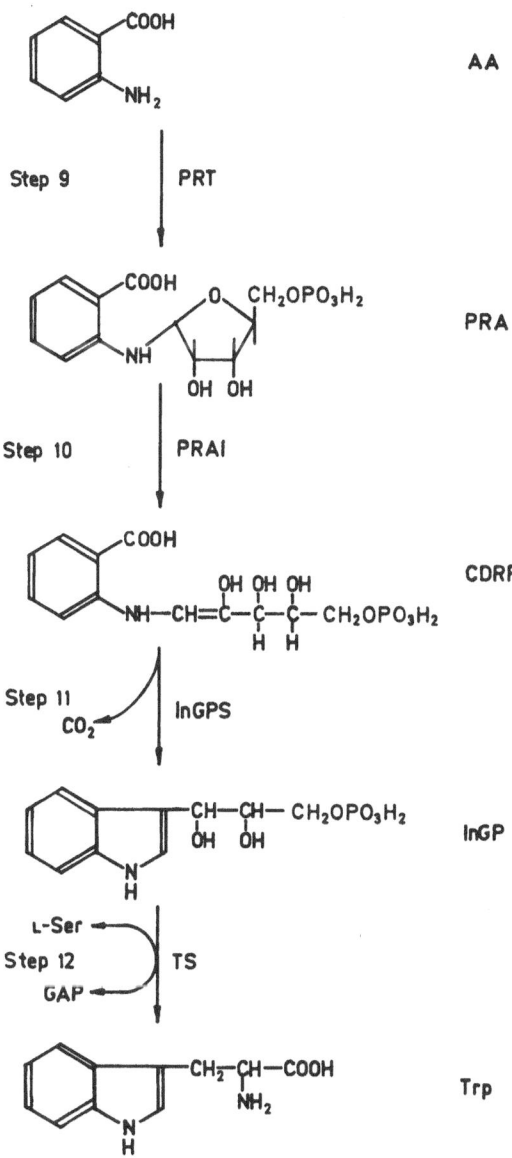

Fig. 2. Tryptophan biosynthesis

The two reaction steps are considered as one enzymatic process, because both are catalysed by one protein aggregate, and free indole cannot be detected in the cells. To become active, this enzyme requires the presence of the coenzyme pyridoxal-5'-phosphate (P5P). Conversion mechanism is partly known. The role of P5P is the binding of serine and its activation in the form of a Schiff's base. By loss of water a double bond is formed in the amino acid, to which indole is associated by its carbon atom in position 3. It has been proved that the amino group of serine is not exchanged during the reaction, while the α-hydrogen attached to the same carbon atom is split off, and another hydrogen atom takes its place [9].

The properties of TS enzyme have been studied in detail on a few microorganisms. Of the procariotic organisms we have a lot of information on the enzyme of *Escherichia coli*, and of the eucariotic organisms the properties of the TS enzyme of *Neurospora crassa* have been studied by several research workers.

In *Escherichia coli*, the enzyme consists of four polypeptide chains, two α and two β chains, which can be easily separated from one another. If the polypeptides occur as $\alpha_2\beta_2$ aggregate, and this is the natural state, both reaction rates are considerably increased. These peptide chains have been very thoroughly investigated, and the amino acid sequence of the two chains is also known [2,6,8,10,11,12].

The TS enzyme of the other microorganism studied in detail, of *Neurospora crassa*, consists also of four polypeptide chains; the aggregate carries the two enzyme activities [8,13,14]. Indole present as intermediate cannot be detected in free form in the liquid; it remains bound to the enzyme between the two reactions [15].

2.4 Degradation and Conversion of Tryptophan

Tryptophan is primarily incorporated into the proteins, this being the main object of its biosynthesis. At the same time living beings possess enzyme systems which break down tryptophan and make it suitable for energy production or anabolism. Moreover, tryptophan is the intermediate of many aromatic, biologically important compounds, which, though very important, are formed only at relatively low concentrations.

In principle all enzymatic reactions are reversible, and thus a backward functioning of tryptophan biosynthesis was also imaginable. However, due to the chemical properties and to the energetic conditions of the given molecules, this cannot be realized. Therefore, a different metabolic pathway was developed in nature, which leads through other intermediates to anthranilic acid or other compounds utilized in metabolism.

Several different bacteria, particularly the facultative anaerobes, thus also *Escherichia coli*, are able to produce indole by the splitting off of the side chain of tryptophan, while pyruvic acid and ammonia is released from the side chain. Indole formation is an important taxonomic mark of bacterial metabolism. P5P coenzyme and K^+ ion (or ammonium ion of about the same size) are needed for the reaction. This enzymatic reaction is reversible if concentration relations are appropriately adjusted. Thus tryptophan is formed from pyruvate, ammonia and indole. This principle forms the basis of several processes for the preparation of tryptophan [16,17,18,19,20,21].

The main pathways of aerobic metabolism is the oxidation of tryptophan [9,19,22,23]. In the reaction, tryptophan-pyrrolase (dioxygenase) enzyme, utilizing molecular oxygen, ruptures the bond between carbon atoms 2 and 3 to form N-formyl-kynurenine. Metabolic pathways converting tryptophan, as shown also in Fig. 3, lead through several branchings to products of different character, including plant and animal hormones (indole acetic acid [24], serotonine), vitamins (nicotinic acid), various substances of dye character (ommochromes, indigo, violaceine, etc.), as well as alkaloids with an indole skeleton (e.g. ergot-alkaloids) [9,25,26,27].

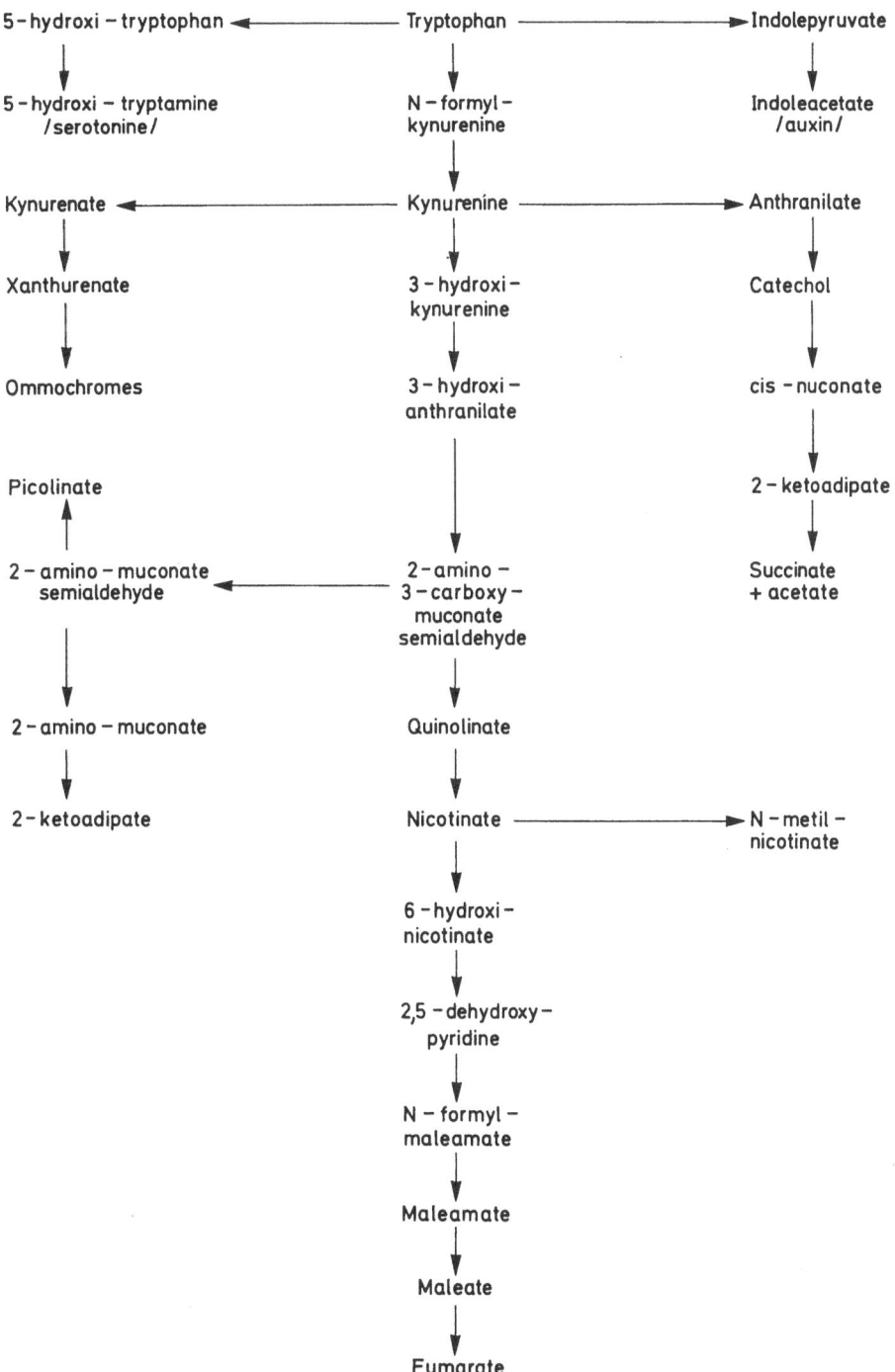

Fig. 3. Conversion metabolites of tryptophan

2.5 Regulation of Tryptophan Biosynthesis

2.5.1 Regulation on Enzymatic Level [7, 28]

DAHP-synthase enzyme, catalysing the first reaction of the tryptophan biosynthetic pathway, is multidirectionally regulated. This step is the first conversion of a branched metabolic pathway; thus, the regulating action of the three end products (Phe, Trp and Tyr), which inhibit overproduction, exerts itself. In the various categories of the phylogenetic system, varied regulation mechanisms have been detected [29, 30].

In the *Enterobacteriaceae* family, e.g. in *Escherichia coli*, the isoenzyme regulation mechanism asserts itself; the single enzymes are under different end-product inhibitions. Of the isoenzymes found in *Escherichia coli* one can be inhibited by Tyr and the second by Phe. The third tryptophan-sensitive isoenzyme is difficult to detect. In the different genera the ratio of the isoenzymes is considerably different [31, 32, 33].

During investigation of 35 strains, representing 21 *Bacillus* species, it was established that, with the exception of one, sequence feed-back inhibition prevails, i.e. there are no isoenzymes, but all three product amino acids diminish the activity of DAHP-synthase. The products inhibit the enzymes of the first reaction step after the branching, which brings about the accumulation of the substrates (chorismic acid, prephenic acid) of these enzymes. The action of these intermediates, present in increased concentrations, is fed back to the enzyme of the first step, to DAHP-synthase [30].

Of the eucariotic fungi, *Saccharomyces cerevisiae* contains two DAHP-synthase isoenzymes, which are Phe- and Tyr-sensitive respectively, Phe reducing the activity by 35% and Tyr by 50%. A third enzyme, or an enzyme fraction under tryptophan product-inhibition, has not been found [28].

Four different isoenzymes of the filamentous fungus *Neurospora crassa* are known. Of these, three are subject to the feedback inhibition of the product amino acids, while the fourth is not sensitive to such actions. This fourth fraction is active even at 45 °C, where the others are already inactivated [28].

This regulation does not act on the next enzymes. Of the variants developed during evolution the most convenient survive, and in the given case the regulation of the first conversion after branching is the most convenient. Accordingly, the enzymes chorismate-mutase and anthranilate-synthase are affected by the product concentration.

Two amino acids (Phe and Tyr) have an effect on chorismate-mutase. The mechanism of this bilateral regulation is different in the various microorganisms. There are examples for isoenzyme regulation, multivalent product inhibition and unelucidated cases alike.

Unequivocal end-product inhibition acts on the enzyme anthranilate synthase; this is a common phenomenon in almost all the living beings. In the presence of tryptophan a significant decrease in activity can be observed, and in several cases even the enzyme of the next reaction, phosphoribosyl-transferase, is partially inhibited [7].

For the anthranilate synthase reaction the amino group is supplied in vivo by glutamine, and this glutamine synthesis affects the tryptophansynthesis.

2.5.2 Supramolecular Organization

It is a feature of the genetic and enzymological background of the tryptophan biosynthesis outlined above that macromolecules (DNA and proteins) reveal various

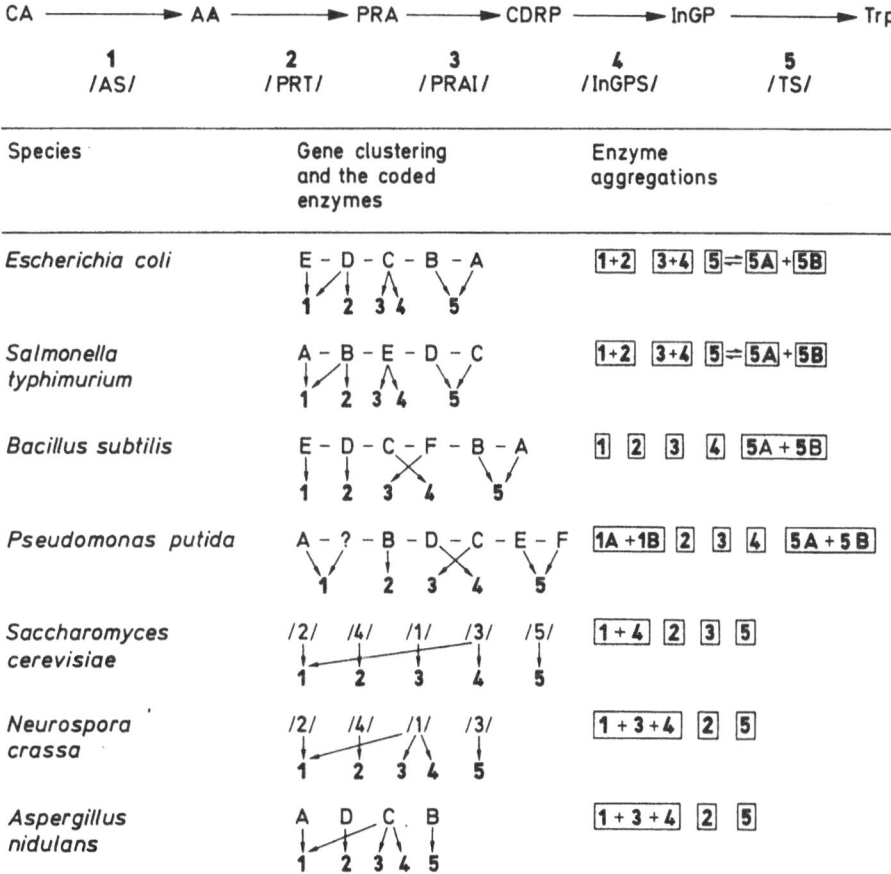

Fig. 4. Supramolecular organization of tryptophan biosynthesis

arrangements in different microorganism species; their mutual order and their grouping have rather changed during evolution. On the basis of publications which use different approaches, we attempt to illustrate the main types using examples [34,7,8]. In Fig. 4 the gene-enzyme relationships and enzyme aggregations of a few characteristic microorganisms are shown.

Procariotic organisms have a relatively simple genetic organization, the genes are aggregated in operons, and accordingly, their regulation is also common [35,36,7].

The genetic structures of *Escherichia coli* and *Salmonella typhimurium* are completely identical, the order of the genes in the operons is the same as the order of the corresponding enzymes in the metabolic pathway. Moreover, the enzymes **1** (AS) and **2** (PRT) form a complex. Only AS in this associated state reveals appropriate activity. The activity of the enzymes **3** (PRAI) and **4** (InGP) is carried by two well-defined sections of one polypeptide chain. The subunits of enzyme **5** (TS) are coded by two separate genes. This arrangement is characteristic of *Enterobacteriaceae* and such was also observed in *Aerobacter aerogenes* [34,7,8].

In *Bacillus subtilis* the enzymes **3** and **4** are coded by different genes, and their order is reversed. Enzyme **1** (AS) appears in the form of two iso-enzymes, and only one of these is located in the tryptophan operon (E-gene). The enzymes of this microorganism do not associate, with the exception of TS(A) and TS(B) proteins, which exert their action jointly [35, 7].

Some bacteria show a more complicated genetic arrangement [22, 7, 37]. Thus, for example, in *Pseudomonas putida* the genes are not to be found at three different places in DNA. The genes of **1**, **2** and **4** enzymes form a close operon-like unit, and show coordination and polarity in the protein synthesis. The first enzyme (AS) consists of two proteins. The genetic location of the peptide of lower molecular mass, which functions as amino donor, is not yet elucidated, and it appears thus as an interrogation mark in the Fig. 4. The C gene (enzyme **3**) belonging to the process, is separated from the others, and is separately transducible. The third genetic group is formed by the genes E and F, coding the two protein chains of enzyme **5** (TS), the genes being located close to each other.

The supramolecular organization of eucarotic microorganisms considerably differs from that of the procariotic; the genes belonging to one metabolic pathway are not aggregated in an operon. The gene maps and the varied patterns of enzyme associations have been well systematized by Hütter and DeMoss in their survey [34]. They classed the forms of organization into five types, and derived on the basis of these a phylogenetic system. We do not wish to present here the whole system, but will describe only two basic types by way of example.

Yeasts are represented in Fig. 4 by *Saccharomyces cerevisiae*. The protein structure of enzymes **1** and **2** is coded by the tr-3 gene, and a single polypeptide chain carries both enzyme activities. For the normal functioning of AS the peptide determined by gene tr-2 is also needed. The genes of the other three enzymes and the enzymes themselves are arranged separately [34, 8].

Neurospora crassa and *Aspergillus nidulans* are examples of a form of organisation often met with fungi. The biosynthesis pathway is stored only at four genetic loci. The enzymes **1**, **3** and **4** form a stable 10 S aggregate, the separation of which yields two protein chains corresponding to the two genes, and the AS activity substantially diminishes, while the other two activities remain unchanged. **2** and **5** are independent enzymes; the genes belonging to them do not interact.

2.5.3 Genetic Regulation

Of the procariotic microorganisms the genetic regulation of *Escherichia coli* and *Salmonella typhimurium* proceeds according to the classical operon-theory: the presence or absence of tryptophan jointly represses or derepresses the formation of all the enzymes of biosynthesis. In other procariotic organisms the genes of the metabolic pathway are located in several groups, which are under separate regulation [7, 37].

Of the eucariotic microorganisms, in the case of *Saccharomyces cerevisiae*, representing the group of yeasts, the formation of anthranilate synthase can be repressed with tryptophan [7, 38].

Of the filamentous fungi the genetic regulation of *Neurospora crassa* is best known. The formation of the enzymes of the synthesis pathway is not coordinated, but is under repression. Regulation proceeds, at least partly, per enzyme, the accumulation of the substrate induces the formation of the given enzyme. This was proved in several works for InGPS and TS. Trp-genes at different loci of the genetic substance are in close connection with the genes of other metabolic pathways. Thus, histidine, arginine and tryptophan biosyntheses are partly under common genetic regulation, histidine or tryptophan defficiency causes the joint derepression of the enzymes of the three metabolic pathways. The regulatory effect of histidine seems to be general in the case of both eucariotic and procariotic organisms (see instances in 3.3.1.) [39, 40, 41, 71].

2.5.4 Effect of Tryptophan-antimetabolites

The antimetabolites of tryptophan are derivatives methylated or fluorinated on the indole skeleton, and indole acrylic acid.

Their action is exerted mainly through the regulating mechanism, which senses also the antimetabolites as real tryptophan concentration, and blocks tryptophan synthesis. In the absence of tryptophan the growth of the cells stops and the culture dies.

The effect of the antimetabolites is utilized in biochemical and metabolic investigations where specific mutants are isolated.

The regulating systems of procariotic organisms are sensitive to antimetabolites. On treatment with antibolites after mutational processes only the resistant mutants begin to grow. One cause of resistance may be the injured regulating system, so that there are mutants which produce constitutively tryptophan without product inhibition. Several tryptophan production processes are based on mutant bacterium strains isolated in this way [37].

Eucariotic cells are less sensitive to tryptophan-analogues than procariotic cells. This is primarily due to selective uptake of substance [42, 43].

In the case of *Saccharomyces cerevisiae*, problems of antimetabolite-sensitivity and resistance were investigated for several compounds. 3- and 4-Methyl-Trp scarcely affect the wild strain at all, but when permeability had been increased the analogue was attached to the tryptophan — selective tRNS-s, thereby slowing down protein synthesis; the analogues led to the stopping of growth. Regulating mechanisms are not involved; the apparent tryptophan deficiency causes derepression of the bio-synthesizing enzymes and intensive tryptophan production. Based on this mode of action, Soviet researchers have patented a process for the production of tryptophan, using 3-Me-Trp [44, 45].

5-Methyl-Trp similarly has a small effect on the wild strain. Its site of attack within the cell is different; it is linked to the product-inhibiting receptor of AS, and blocks tryptophan synthesis. It has no effect on genetic regulation, because the enzymes of the synthesis pathway are derepressed in its presence.

The mode of action of 5-F-Trp shows a further variation: it is incorporated into the proteins, and produces thereby proteins with faulty function. It acts strongly also on the wild strain since the permeability barrier is no obstacle. Moreover, it causes product-inhibition on AS, reducing its activity, and thus inhibits also the synthesis of tryptophan.

5-Fluoro-indole penetrates the cell either by physical dissolution or diffusion. In the cell interior it is converted by the TS(B) enzyme into 5-fluoro-tryptophan. Following this, its mode of action is identical with that of 5-F-Trp [42, 43].

Indole acrylic acid can also be classed among the antimetabolites. It is linked to the TS(B) enzyme and paralyses its action. Growth stops when tryptophan is lacking, while the response of regulating mechanisms is derepression, which causes high enzyme production and InGP accumulation. This general derepression and accumulation of InGP induce also the TS enzyme [39, 42].

3 Production of Tryptophan

Several possibilities are offered for the production of tryptophan. A method interesting only from the historical aspect is recovery from protein hydrolysates. Under appropriate conditions, organic syntheses, enzymatic and microbial processes can be used even today.

The main object of this review is the discussion of microbial processes; synthetic and enzymatic processes will be only briefly summarized.

3.1 Chemical Synthesis and Resolution Methods

About twenty synthesis variants are known for the preparation of L-tryptophan, of which a very good survey was published by Kaneo and Izumi [46]. Processes can be grouped according to the following:
a) from compounds with indole skeleton:
    aa) from indole-3-aldehyde (e.g. with hippuric acid)
    ab) from gramine (e.g. malonic ester, acetoacetic ester syntheses)
    ac) reaction routes starting from indole
b) total syntheses by the building up of the indole skeleton.

All the organic syntheses yield DL-tryptophan, the racemic mixture. The product obtained must be resolved for the separation of the biologically active L-isomer. Indeed, the economy of a synthetic technology depends in many cases on the efficiency of resolution.

For the separation of the tryptophan stereoisomers the fractional crystallization of the zinc or ammonium salt of the N-acetyl derivative is recommended. It can also be isolated in the form of its complex with inosine [46].

In addition to chemical methods, enzymatic hydrolysis is widely used. When the DL-amino acid derivation prepared by synthesis is treated with the appropriate hydrolases, only the L-isomer is liberated. Substrates of the following types are generally used:

> DL-amino acid esters
> DL-amino acid amides
> N-acyl-DL-amino acids
> DL-amino acid hydantoins.

Several publications deal with the resolution of tryptophan on this principle. Japanese authors have reported on the hydrolysis of synthesized N-acyl-DL-trypto-phan with aminoacylase of mould origin [46]. According to another Japanese process, the hydrolysis of DL-tryptophan-hydantoin is carried out with a microorganism strain isolated from soil. No enzyme preparation was made of the cells, the racemic substance was added to a resting cell culture [47, 48].

Okazaki isolated a *Pseudomonas* strain able to isomerize tryptophan, with which L-tryptophan can be prepared from D-tryptophan in a resting cell culture. For the reaction another cheaper L-amino acid suitable for transamination is needed [49]. An essential part of all the resolution processes is the racemization by heating or treating the residual D-amino acid or D-amino acid derivative with alkali, and the recirculation of the racemic mixture into the process.

3.2 Enzymatic Processes

Knowing the metabolic pathways, tryptophan can be formed in enzymatic reactions of two kinds. The TS enzyme produces tryptophan from indole-3-glycerol-phosphate, while tryptophanase from indole, pyruvic acid and ammonium ion. In consideration of the fact that the enzyme TS(B) is also able to produce tryptophane from indole and serine, it becomes evident that tryptophan can be obtained in two ways from indole.

Of the further substrates needed for the reaction, L-serine is the most expensive, because here too a racemic mixture is formed during manufacture. Thus preferably the cheaper raw materials, pyruvate and ammonia, are used. This subject matter was summarized by Abbott [16] and Esaki et al. [3].

Italian researchers used *Escherichia coli* enzyme, chemically binding the enzyme to macromolecular fibres [11].

The patents of the Ajinomoto Co. concern the production of tryptophan and 5-hydroxy-tryptophan using preparations made of *Proteus*, *Erwinia* and *Escherichia* strains and resting cells [50, 18]. Though they call the enzyme used by them tryptophanase, the facts that the other substrate used besides indole is serine or β-chloroalanine, and furthermore that the ammonium ion is not present in the reaction mixture in all the examples, indicate that they actually work with a TS enzyme preparation.

In the other enzymatic reaction, the tryptophanase enzyme is most often obtained from *Escherichia coli* or from the biomass of the genera *Proteus*. In cultivation, the synthesis of the enzyme can be induced with tryptophan. Tryptophanase of *Escherichia coli* is stable and relatively easy to isolate. The presence of P5P and K^+ ions are needed for the reaction. P5P is chemically bound to the enzyme, and through it the enzyme can be linked to macromolecules, e.g. to cellulose triacetate. The functioning of the enzyme can be well described by a kinetic equation valid for three substrates. Indole at a concentration higher than 0.1 mM causes substrate inhibition. These processes have been described only on a laboratory level; pilot plant or industrial manufacture has not been reported [16, 12, 20, 52, 53, 54, 3].

In processes using tryptophanase prepared with various *Proteus* strains, the enzyme is often not isolated, but introduced as a "resting cell" into the reaction mixture. In this case too, P5P and K^+ (or NH_4^+) ions are needed for the functioning of the enzyme [17, 21].

3.3 Production Processes

The preparation of L-tryptophan by a bioprocess also yields pure L-tryptophan, so that the costly series of resolution operations is omitted.

Two types of microbial processes, differing in principle, can be distinguished with respect to tryptophan production:
a) "de novo" syntheses without precursor,
b) bioconversions from tryptophan-precursors.

3.3.1 Processes Without Precursor

Various microorganisms continuously produce tryptophan, but overproduction is prevented by the multiple efficient regulation systems discussed earlier. It is the task of the microbiologists to shift bioregulation in a direction where the micro-organisms produce a high quantity of tryptophan without any feedback.

Such changes were brought about by induced mutation in several microorganism strains. Mutants isolated by different techniques are mainly either aromatic amino acid auxotrophic and/or antimetabolite-resistent mutants. The simplicity of genetic regulation as well as partly in conjunction with this, the antimetabolic-sensivity are characteristic of procariotic microorganisms. In microbial processes without a precursor almost exclusively mutant bacterium strains are used (Table 1).

Kida and Matsushiro prepared first from *Escherichia coli* of well-known regulation a mutant with injured feedback both on the enzyme level and gene

Table 1. Tryptophan production by antimetabolite resistant mutants

Strain	Antimetabolite resistance	Trp accumulation g l^{-1}	Other characteristics	Ref.
Bacillus subtilis	5-Me-Trp	—		[55]
Bacillus subtilis GEN-37	5-Me-Trp	6.7–10.1		[56, 70]
Bacillus subtilis K	5-F-Trp	6.10	Leu-bradytrophic	[57]
Bacillus subtilis S-10	5-Me-Trp	11	anthranilic acid, indole precursor	[58]
Azotobacter suis	3-Me-Trp	3.5– 4.5	Tyr auxotrophic	[44]
Escherichia coli	5-Me-Trp	2.5	Trp auxotrophic and anthranilic acid resistent	[59, 51]
Escherichia coli K 12	5-Me-Trp 5-F-Trp	0.025		[60]
Brevibacterium flavum	5-F-Trp	6.2–12	3-F-Rhe^R, His$^-$, Phe$^-$	[61]
Corinebacterium glu-tamicum ATCC 21851	4-Me-Trp, 6-F-Trp	11.5–13.8	Phe$^-$, Tyr$^-$, 4-NH$_2$-Phe^R, 4-F-Phe^R Tyrhx^R, Phehx^R β-thienyl-Ala^R,	[62, 63, 64, 65]
Methylomonas metanophyla	5-Me-Trp	0.2	3-NH$_2$-Tyr^R, methanol substrate	[66]

levels [67, 68]. This strain produced on the usual culture medium 0.4 g l^{-1} tryptophan, while in an optimized culture medium complemented with anthranilic acid an enhanced yield of 0.7 g l^{-1} was obtained.

Several researchers isolated independently of one another well producing mutants also from *Bacillus subtilis*. A multistep mutational process is described by Shiio et al.: after the investigation of several thousand resistant mutants a leucine bradytroph colony resistent to 5-fluorotryptophan was found, which produced in 72 h 6.15 g l^{-1} of tryptophan. The linkage of the regulating systems is proved by their finding that their resistant and producing mutant was overproducing also with respect to phenylalanine. An other *Bacillus subtilis* mutant of 5-fluorotryptophan resistant but leucine auxotroph also produced tryptophan (5.3 g l^{-1} in 72 h). This effect of leucine auxotrophy or bradytrophy is interesting because it contradicts the fact that leucine generally activates the enzyme prephenate dehydratase and reverses the inhibition by tryptophan [57].

The *Bacillus subtilis* strain isolated in the USSR after multiple mutations is phenylalanine auxotrophic and 5-methyltryptophan resistant. The strain produces a maximum of 10.1 g l^{-1} of tryptophan [56, 69, 70] in 72 h.

This product concentration approaches already the solubility limit of tryptophan in various culture media (12–15 g l^{-1}). Phenylalanine and tyrosine auxotrophy are very important in these cases, because the faults of biosynthesis prevent the transformation of the common metabolites into phenylalanine and tyrosine, so that the whole quantity of chorismic acid is converted into tryptophan. Petrov et al. prepared and isolated tyrosine-auxotrophic mutants from a facultative anaerob nitrogen-fixing strain of *Azotobacter suis*. The best mutants produced 0.2–0.8 g l^{-1} of tryptophan, however, when a small quantity of 3-methyl-tryptophan was repeatedly added to the culture medium during cultivation, production increased to 3.5–4.5 g l^{-1} [44, 45]. This phenomenon can be well interpreted by antimetabolite-resistance, because 3- and 4-methyltryptophan, bound to tryptophan specific tRNAs produces an apparent tryptophan deficiency, and thereby enchances enzyme synthesis.

Some microbial processes using hydrocarbons as sole carbon-energy source have also been elaborated. Moran and Raboin [71] isolated *Mycobacterium rhodochrous* 5-methyltryptophan resistant, histidine-, tyrosine- and phenylalanine-auxotrophic mutants able to produce tryptophan. The accumulation is but 0.55–0.77 g l^{-1}, the C-source is n-paraffin from C_8 to C_{20}. Investigations of Japanese researchers, in which various His$^-$ mutants were isolated, yielded similar results. They achieved with their *Arthrobacter paraffineus* His$^-$ mutant, grown on both sorbirol and kerosene, in 96 h a product concentration of 0.4–0.5 g l^{-1} [62]. The antimetabolite-resistent mutant of *Arthrobacter carbazolum* yielded (in a culture medium containing 1 % of carbazole, an indol-precursor or -derivative) a product concentration of 2 g l^{-1} [72].

Suzuki et al. produced on methanol carbon source all the three aromatic amino acids, after the isolation of mutants resistant to the antimetabolites of the three products [66].

The Japanese industry realized tryptophan production with mutants obtained from *Brevibacterium flavum* (Ajinomoto Co.) and from *Corynebacterium glutamicum* (Kyowa Hakko Kogyo Co.) strains. Shiio et al. arrived from the wild type strain of *Brevibacterium flavum* with purposeful, multistep mutational work to the strain of good production capacity. In the first phase they endeavoured to isolate the $5MT^R$, Tyr^-, Phe^- mutants. Next, the effect of all the three aromatic amino acids on the branching steps and on the DAHPS enzyme was suspended.

To achieve this, phenylalanine antimetabolite (3FP) was also used. The strain meeting the demands ($5MT^R$, $3FP^R$, His^-, Phe^-) accumulated 6.6 g l^{-1} tryptophan in the medium (His-deficiency probably causes derepression of the enzymes of the tryptophan biosynthesis as in the case of *Neurospora*, see Ref. [40]). By the further development of the process and its optimization at intensive aeration, 12 g l^{-1} concentration was attained in 56 h [65,61,73].

Nakayama et al. worked on similar principles in the isolation of the mutants of *Corynebacterium glutamicum*. The strain which proved to be the best producer possesses the following properties:

Phe^- and Tyr^-

$5MT^R$ and $6FT^R$

4-amino-Phe^R and 4-F-Phe^R

Tyr-hydroxamate^R and Phe-hydroxamate^R

These properties mean the ceasing of all negative feedback affecting the metabolic pathway. Using high sugar concentrations (about 15%) during cultivation, 16.8 g l^{-1} of tryptophan was accumulated in the medium after 72 h. A modified process was elaborated with the same strain on ethanol instead of sugar or molasses as the carbon source. During cultivation the ethyl alcohol (about 12%) to be introduced is fed in at a rate which maintains a concentration of between 0.5 and 1.0% [63,64,65].

Tribe and Pittard reported a more elaborated genetic manipulation on *Escherichia coli*, which resulted in a lot of genetically modified mutants capable of hyperproduction of tryptophan. One of these regulatory mutants was able to produce about 1.3 g l^{-1} tryptophan in 25 h in a 1.0 l laboratory fermenter with a specific tryptophan production rate as high as 80 mg per g dry weight per hour [116].

3.3.2 Precursor Procedures

In microbial processes belonging to this category the changed metabolism regulating system is of less importance than in "de novo" biosyntheses.

Here the task is not the enhancement of the activity of the whole metabolic pathway, but "only" the acceleration of the reaction steps between precursor and product, the suspension of inhibition.

Accordingly the sphere of cultured microorganisms is not limited to procariotic organisms, but processes utilizing eucariotic strains have also been developed. Right at the beginning, a strain of *Claviceps purpurea* was isolated from the wild strains in the Eli Lilly laboratory; it was able to synthesize large quantities of tryptophan from indole. The glucose, $(NH_4)_2SO_4$ and corn steep liquor contents of the culture medium were optimized. The pH of the medium was not controlled, the final product concentration was $1.5-1.6 \text{ g l}^{-1}$ [74,75].

Mateles and Lim used 5MT- and anthranilic acid-resistant *Escherichia coli* strains. In this case anthranilic acid-resistance means that even at higher concentrations the cells tolerate this aromatic metabolite, which has a growth-inhibiting effect on wild strains. Owing to mutation, one of the enzymes of the strain became inactive in the metabolic pathway before anthranilic acid. Since this derepresses the further biosynthesizing enzymes, anthranilic acid added to the culture is converted into tryptophan. The production capacity of the strains is low, about 0.2 g l^{-1} [51,59].

A product of Japanese fermentation research is the technology based on an anthranilic acid-resistent and anthranilic acid-auxotrophic mutant isolated from *Bacillus subtilis*. The genetic injury is similar to that of the *Escherichia coli* mutant mentioned in the aforesaid, the microorganism efficiently transforms anthranilic acid into tryptophan. In addition to the usual carbohydrates, dicarbonic acids (succinate, tartarate) are also added as carbon sources to the culture medium. During batch cultivation 10–12 g l^{-1} of tryptophan is formed in 48 h [76]. The brief fermentation time may counterbalance the costs of the precursor, and the procedure could thus possibly compete with "de novo" processes.

Czecho-Slovakian researchers produce tryptophan with a *Corynobacterium* sp. strain of high indole tolerance. They use a very high (20%) sugar concentration in the culture medium, and in 50 h about 10 g l^{-1} of tryptophan is formed [77,78].

Some processes realize tryptophan production on a methanol carbon source. They culture methanol-utilizing microorganisms, which metabolize C_1 compounds via the serine pathway. Serine is formed as an intermediate, and can be diverted in the presence of adequate tryptophan-synthase enzyme activity and indole concentration in the direction of tryptophan production [79]. The strains used in the patented process of the Swedish Bofors AB were selected from the *Pseudomonas* and *Methanomonas* genera. Besides methanol and indole, the culture medium contained only mineral salts. The toxicity limit for methanol was 5%, for indole 500 mg l^{-1}. The culture was kept under oxygen limit, because the decomposition of the product is slower under these conditions. In batch culture, a product level of 3.3 g l^{-1} was achieved with the best strain in 48 h [80].

Using *Brevibacterium ketoglutaricum*, *Candida tropicalis* and *Streptomyces anti-bioticus* cultures, Japanese researchers converted anthranilic acid into tryptophan. The best strains, cultured on n-paraffins (C_{11} — C_{18} fraction), were able to accumulate 0.8–2.0 g l^{-1} in 72 hours [81]. Owing to the low product concentration and to the very high use (20 g l^{-1}) of precursor, the process does not seem economical.

3.3.2.1 Realization of Bioconversions with Yeast Strains

Precursor processes carried out with yeasts are discussed separately, because such processes were developed in several research laboratories, independently from one another, on the basis of the same research principles. So far, four research laboratories have published their results:

1) In the USSR, the Microbiological Institute "Kirszenstein" of the Lithuanian Academy of Science, with *Candida utilis* 295 t yeast strain.
2) In Japan a process was elaborated for Kyowa Hakko Co. using *Hansenula anomala* I-22 yeast strain.
3) In Sweden commisioned by Bofors AB, with *Candida humicola* yeast strain.

4) In Hungary, the Technical University of Budapest, with *Hansenula anomala* yeast
   strain.

3.3.2.1.1 Origin of the Strains

In all the cases experiments were begun with wild-type strains, among which
microorganisms able to produce tryptophan were found. The starting point of the
Soviet researchers was to increase the tryptophan content of common yeasts
grown on molasses by the addition of anthranilic acid. Anthranilic acid added in
small concentration (max. 0.05%) increased the quantity of free tryptophan in the
cells as well as that bound in proteins. Increase was highest with the yeast
Candida utilis [82]. On the basis of this, in further experiments aimed at producing
pure tryptophan, the microorganism *Candida utilis* 295t was used, which can be
considered as a wild type strain. Mutation experiments were also performed, but the
results did not prove applicable [83].

For the preparation of tryptophan on anthranilic acid basis, Terui et al. [84] under-
took the purposeful screening of strain collections and of isolated strains of natural
origin. After the investigation of several hundred strains they established in general
that wild type bacteria are less suitable for this conversion, while among yeasts the
species of *Candida* and *Hansenula* genera revealed important activity. Research
results were patented already at this stage; the strain found best, *Hansenula anomala*
produced $1.2 \, g \, l^{-1}$ of tryptophan [85]. The production capacity of the strain was
increased by mutational methods. Mutational techniques were complicated by the
circumstance that the microorganism is of diploid type, and thus it is more difficult to
produce and preserve genetic changes. They used a modified variant of Wickerham's
method and, in mutant isolation, took into consideration the following aspects as
prerequisites of high tryptophan production: indole-tolerance, rapid growth at
pH 4, toleration of $5 \, g \, l^{-1}$ anthranilic acid at pH 4.

Of the mutants prepared that of highest production capacity produced $6 \, g \, l^{-1}$
tryptophan according to the standard method of testing. Further development work
was undertaken with this strain [86,87].

Swedish researchers do not report in detail on strain selection, but communicate
only that the strain isolated from nature (from soil) was identified as *Candida
humicola* and thus can be considered as a wild strain. They add that there is no
kind of product regulation and in batch culture the microorganism achieves a product
level as high as $15 \, g \, l^{-1}$ from indole. The strain is genetically stable; even during
long-time cultivation no deterioration, degeneration was met [88].

After systematic screening of wild type strains, Hungarian researchers selected
the *Hansenula anomala* strain. With indole precursor, the strain accumulated $12 \, g \, l^{-1}$
tryptophan according to the standard testing method [89,90,91,92]

3.3.2.1.2 Composition of the Culture Medium

In all cases, the experiments were begun on culture media containing carbohydrate,
but alcohol added with the precursor was also utilized as a carbon source.

Soviet researchers attempted to improve the composition of the culture medium,
though they did not apply scientific methods for this purpose. They selected only
those media from the arbitrarily adjusted nutrient solutions which gave the best

production. The culture medium that proved best contained 200 g l^{-1} molasses and 7.5 g l^{-1} urea. Moreover, they investigated the effect of the addition rate of sugar and anthranilic acid [93, 94].

The Japanese researchers carried out qualitative and quantitative culture medium experiments to elucidate the nutrient requirement of the *Hansenula anomala* strain. According to their measurements glucose, saccharose and glycerol can be used with good results as a carbon source, while ammonium nitrate and urea can be taken as nitrogen sources. They also investigated the effect of the trace element composition of the culture medium, and gave the optimal Mg^{2+}, Fe^{2+} and Zn^{2+} concentration. The optimal value of dissolved oxygen was found to be about 1.3 mg l^{-1} [95, 96]. On comparing the three carbon sources mentioned, the utilization with respect to the product increased in the order ethanol, glucose, glycerol [97].

Swedish researchers only gave the composition which they found to be optimal. The carbon source was glucose, and particular attention was paid to the iron (II) ion concentration, which plays a role in the decomposition of tryptophan. For the very reason of its high iron content, they did not use molasses for culturing. They regulated the dissolved oxygen concentration and iron ion concentration to repress the decomposition. At high iron concentrations they adjusted low dissolved oxygen concentrations, and vice versa [88]. The culture medium did not contain other organic substances besides carbohydrate, thiamine and niacin.

In the Hungarian process, the carbon source is ethanol. On comparing the carbon and nitrogen sources, it has been established that ethanol, saccharose and molasses are the best carbon sources, while ammonium sulfate and ammonium chloride are the best nitrogen sources. Of the growth factors, biotin and thiamine were added to the culture medium. The most favourable nitrogen, phosphate and biotin concentrations were determined by the Box-Wilson optimization method [90, 92].

As all the processes require a long time, nutrients must be periodically or continuously added during the process. This means on the one hand the addition of alcohol, as precursors are always introduced in concentrated alcohol solution, while on the other hand the feeding in of sugar and mineral salts is also general practice.

3.3.2.1.3 Technological Aspects

The processes discussed above have several common features. Intensive aeration needed for yeast propagation is characteristic of all of them; high cell concentration must be attained for product formation. An important part of the substrate used is expended on growth and energy production, and only a smaller part for the transformation of the precursor. With yeasts, the optimal pH value generally lies in the acid region; nevertheless these processes are carried out in a nearly neutral medium, because this range is more favourable for tryptophan biosynthesis. A common feature is the long cultivation time, and because of this, the periodic introduction of various ingredients (alkali, precursor, nutrients) is also necessary in all the technologies.

Fermentation technology based on *Candida utilis* has been fundamentally divided into two parts. In the first period, the aim is the maximal production of the biomass, and fermentation parameters are adjusted accordingly. In the interests of rapid growth, intensive aeration was applied; sulfite numbers of 200–280 mmol O_2 $l^{-1} \cdot h^{-1}$ were reported, which have been achieved in a bioreactor of special design, under application of overpressure. As unusual with yeasts, the pH was adjusted between 7.5 and 8.0 during cultivation. The growth period lasts about 24 h. In the production period, continuous tryptophan production was ensured by the periodic addition of sugar and anthranilic acid. The optimal aeration is lower than before, according to experimental results it can be characterized with a sulfite number of 120 to 150 mmol O_2 l^{-1} h^{-1}. The duration of the productive phase is 144 h, (pH 7.5–8.0, temperature 28–30 °C), the same as the values adjusted in the growth period. From the 4.2 g l^{-1} of anthranilic acid fed into the process 6.4 g l^{-1} tryptophan is formed [98, 99, 100, 101, 102, 103, 104].

In the process with *Hansenula anomala*, developed by the Japanese researchers, technological parameters do not change during the whole process; batch cultivation was carried out at 27 °C, in a medium buffered with $CaCO_3$, in a 30 l bioreactor agitated at a speed of 500 rpm, and applying an aeration of 2 vvm. The starting volume was only 12 l, but this was later considerably increased by the added ingredients. The activity of tryptophan-biosynthetic enzymes was investigated as a function of time. It was found that the specific product formation rate gives a maximum curve and all the measured enzyme activities reach the maximum value at the same time. The maxima of the specific product formation rate and of the specific growth rate do not coincide; product formation lags behind by about 10 h as compared to growth [105]. 14 g l^{-1} of tryptophan was formed in 250 h. Besides the addition of the carbon source, the long reaction time made necessary the addition of all the mineral salts [106].

The Swedish technology is based on a continuous process. There are no data on production on a larger scale. Technological parameters are similar to those mentioned above: The reaction was carried out at 26 °C and pH 6.5. The growth cycle was about 24 h, then cultivation was made continuous. The dilution rate was about 1 day^{-1}. Aeration was accomplished at a very high speed of rotation (2000 rpm, in a 10 l reactor) and at a relatively low air introduction (0.2–1.2 vvm). The decomposition of tryptophan, or rather the prevention of decomposition plays an important part in the process. On samples systematically taken the concentration of kynurenic acid (decomposition product of tryptophan) was determined by its UV absorbance, and was reduced to a minimum by the regulation the concentrations of iron ions and dissolved oxygen. The tryptophan content in the fermentation broth drawn off attained values as high as 10.4 g l^{-1} [88].

The Hungarian process is a fed batch fermentation. During the process, ethanol is fed continuously on the basis of the control signal of the dissolved oxygen concentration. For nitrogen supply and pH control, ammonium hydroxide solution is added. The culture grows in oxygen limit, which results in a linear growth curve. The increase in tryptophan concentration is also linear during the growth. After 120 h, the culture medium contained about 12 g l^{-1} of tryptophan and about 80 g l^{-1} of dry biomass [89, 90, 91, 92].

3.3.2.1.4 Precursors, Feeding of Precursors

Both indole and anthranilic acid injure living cells above a certain concentration. This injurious limit value is rather high for fairly tolerant yeasts, being about

200 mg l^{-1} in the case of indole, and about 1 g l^{-1} in the case of anthranilic acid. Knowing this, the total precursor quantity cannot be introduced on one occasion into the system. Repeated or continuous feeding is needed, which is regulated on the basis of the actual precursor concentration measurements. It is noteworthy that yeast cells, even dead ones, bind a considerable quantity of indole (3.5–4.0 mg per g dry biomass). The process is possibly of a partition character; the cell membranes, rich in lipids, effectively extract indole from the aqueous phase [106].

In three processes, the feeding of the precursor has been linked with the addition of the supplementary carbon source. Sugar solution was fed each 3–6 h into the C. utilis culture and, simultaneously or with some delay, about 0.3 % of anthranilic acid in alcohol solution.

The Japanese researchers have developed a very convenient method of feeding. When the energy source, be it sugar, glycerol or alcohol, was exhausted in the medium, the microorganisms, with its decreasing substrate consumption, also reduced its oxygen consumption. However, at unchanged aeration intensity, this resulted in an increase of the dissolved oxygen concentration. Under continuous measurement of the dissolved oxygen concentration, a control loop was designed, which feeds in substrate and precursor (e.g. indole solution in alcohol), whenever the concentration of dissolved oxygen increases [106]. This effect, namely the steep descent of the dissolved oxygen concentration caused by the feed, could be observed also on the record published in the Soviet paper [99]; however, they gave no interpretation. In the investigation of Hansenula anomala, experiments were carried out with the feeding in of both precursors. For a long time, the object of their research work was the transformation of anthranilic acid, and only when a suitable microorganism and small-scale technology were available for this purpose, did they begin with the feeding in of indole. In the first growth period, the medium contained only anthranilic acid, and a further feeding in of precursors was begun only after 12–24 h. Indole was introduced as alcohol solution, and its feeding was regulated so that at each occasion only about 100 mg l^{-1} of indole was introduced into the system. The level of anthranilic acid was raised by each addition to about 1.0 g l^{-1}. This double precursor feeding accelerated fermentation. The same strain of yeast also proved to be a good producer with indole as sole precursor. By feeding indole solution in ethanol into sugar-containing basic culture broth, 10 g l^{-1} of product were accumulated in 72 h [107].

The Swedish continuous technology uses indole exclusively as precursor. Indole was fed in independently of the culture broth additions, so as to maintain indole concentration at a constant value (about 200 mg l^{-1}). Besides indole, the biosynthesis of tryptophan also requires serine. When serine was added to the culture, it accelerated tryptophan production. However, since serine is too expensive an ingredient, glycine, an amino acid biochemically closely related and easy to transform, has been tested. Glycine accelerated the process in the same way, but only with a few hours of delay [88].

Hungarian researchers achieved the feeding in of ethanol analogous to the Japanese process mentioned, but used separate precursor addition on the basis of their experiments. According to the measurement of indole concentration, they maintained the approximately 200 mg l^{-1} level of indole with a 10% ethanolic indole feed solution [89, 92].

3.4 Preparation of Substituted Tryptophan Derivatives

Tryptophan derivatives substituted on the aromatic skeleton are important for the pharmaceutical industry because they are the basic substances of the synthesis of several active substances, e.g. serotonin. 6-Chlorotryptophan can come into consideration as sweetener. Manufacture is aimed at the production of substances small in quantity but high in purity [108, 109].

In principle, the methods of preparation surveyed so far also suitable for the preparation of substituted products, but with biological systems the following interfering effects must be taken into consideration:
1) the strongly substrate-specific binding sites of the enzymes are not able to bind all the molecules with substituted indole skeleton;
2) the substituted tryptophan derivative formed by biosynthesis can behave as an antimetabolite and may destroy the cells.

Considering these facts, experiments for the preparation of substituted tryptophan derivatives were aimed at the trasnformation of substituted indole by enzymatic or precursor fermentation technology. A substantial part of the enzymatic processes can be also applied to substituted derivatives. Thus, using *Escherichia coli*, Watanabe and Snell [20] prepare with 5-hydroxytryptophan tryptophanase from 5-hydroxyindole. Sano and Mitsugi [18] carried out the same transformation, using 5-hydroxyindole and β-halogen alanine substrates and *Proteus morganii* enzyme. Tryptophanase, prepared by Hitoshi et al. [17] from *Proteus mirabilis*, has a broader transformation spectrum. The reaction proceeded also with derivatives substituted in position 5 by hydroxy-, amino- and methyl-groups.

Of the bioconversion processes too, the most intensively studied field is the 5-hydroxyindole → 5-hydroxy tryptophan transformation. Several strains patented for tryptophan production are also suitable for the preparation of 5-hydroxy-tryptophan. The tryptophan-producing microorganism isolated from *Claviceps purpurea* in the Eli Lilly Laboratories was tested 12 years after it was first described for the conversion of various analogues. From 2-, 4-, 5-, 6- and 7-methyl-indole and from 6-chloro-indole the respective derivatives were formed, while 1-methyl, 6-trifluoromethyl, 6-nitro and 4-benzyloxy-indole were not converted [110].

Besides producing tryptophan the yeast strain *Candida humicola* is also able to synthesize substituted derivatives with hydroxy, halogen, alkoxy and alkyl groups in position 5 [88].

4 Isolation of Tryptophan

Two ways are offered for the isolation and purification produce. On the one hand, the preparation of crystalline tryptophan for the chemical industry and for parenteral solutions, and on the other hand, concentration of the culture medium for nutrition purposes, so as to obtain a feed yeast mass of high tryptophan content, which can be directly mixed with the fodder and fed to the animals.

The first step in the production of pure tryptophan is in all the cases the removal of cells. According to the character of the microorganism, this is realized by filtration or centrifuging. Following this, the methods described in the literature are of two kinds. One part of them binds tryptophan on an ion exchange resin, while others use activated carbon.

It is well known that in the sorption of amino acids on ion exchangers both the amino and carboxy groups and the apolar part of the compound participate. Though the sorption of the amino acids on sulfonic acid resins in the hydrogen form is salt formation, at the same time some kind of "physical adsorption" must also be taken into consideration [111,112].

Other parameters besides pH also affect the sorption of tryptophan. Thus aromatic amino acids can be readily eluted with solvent mixtures, e.g. with ammonium acetate buffer of pH 4.5, containing 40% of alcohol. In the separation of basic amino acids on a sulfonic acid resin equilibrated with a buffer of pH 5, eluting by increasing the sodium ion concentration, tryptophan appeared separated from the zone of neutral and acidic amino acids, but before the basic amino acids.

Researchers engaged in the preparation of tryptophan used various types of cation exchange resins in the hydrogen form, e.g. Dowex 50 [107,81,88], KU-2 resin of Soviet manufacture [99,113,101], Amberlite IR-118 [11], and Diaion SK 104 [65]. As the molecular size of tryptophan is fairly large, it can penetrate resins with dense crosslinking to a lesser extent. Therefore, Paulsson et al. [88] recommend the use of polymers with a lower degree of crosslinking.

For elution from the ion exchange resins alkaline solutions are used. The eluent generally used is 0.3–5.0% ammonium hydroxide; for example Paulsson et al. use it as a solution in 50% ethyl alcohol [88].

In conjunction with sorption on activated carbons, the character and the quality of the carbons are not given in the publications, and thus concrete data are not available.

After the filtering off and washing of the activated carbon, elution is carried out with solvent systems, which can be classed into two basic types. The eluent is either 50–100% alcohol [76,96], or a 4:1:1 mixture of n-butanol, water and ethyl alcohol, the pH of which is adjusted by ammonium hydroxide to 9.6 [74,75,110].

Independently of the mode of binding and elution, the next step of processing is evaporation. This is generally carried out at reduced pressure, at 45–60 °C to provide for efficient crystallization; the pH is adjusted to the isoelectric point. The raw product obtained is purified by recrystallization, using 50–80% ethyl alcohol as solvent.

For the separation of substituted tryptophan derivatives an Avicel (cellulose) columns is used. The eluent phase is a 6:1 mixture of acetonitrile-water [110].

The other end-product of processing may be a biomass with considerably higher tryptophan content than usual. Similar to feed yeast, this is used as direct fodder. The aim of certain researchers is only to increase the intracellular tryptophan content in feed yeasts [82,114]. In other processes, the economic efficiency of manufacture is to be increased by also utilizing the biomass formed as a by-product, which has a high protein and tryptophan content [115,68,113].

5 Present State of the Production and Use of Tryptophan

The yearly production of tryptophan varies between 50 and 100 tons. This refers to the L-isomer, DL-isomers are sold in similar quantities [108, 5, 4].

Even under unchanged conditions and even if tryptophan is no longer used as fodder additive, market estimations predict a marked increase in demand.

The tryptophan market is dominated by Japanese companies (Ajinomoto, Kyowa Hakko, Tanabe); they are responsible for almost the total world production. The companies own several manufacturing processes (synthetic, enzymatic, fermentation and combined processes), but presumably synthetic or semi-synthetic processes are used [5]. In 1977 the marketing price was 120–140 $ per kg but since then this price has considerably decreased, and in 1979 was about 40–50 $ per kg, but now it is again between 100–120 $ per kg.

The spectrum of consumption in kilograms is shown in Table 2 [108].

Tryptophan is not used at present as a fodder supplement because soya, the number one tryptophan source, can be more cheaply purchased. Of the amino acids limiting increase in weight, tryptophan stands at third to fourth place, lysine and methionine unequivocally taking precedence. Thus, for the time being less attention is paid to the tryptophan content of fodder.

The price of tryptophan in soybean was 16–24 $ per kg, as compared to the prices mentioned above. According to theoretical calculations, a plant could be set up which would economically manufacture tryptophan even at this low price, but only if it works at a very high capacity (about 1000 t per year); this is a multifold of the present world production. In countries where soya as a fodder component is difficult to obtain, the production and use of tryptophan may be justified.

The above market research data do not include the production and consumption of the socialist countries. In the Comecon countries, tryptophan is manufactured at present only in small quantities for fine chemicals.

The facts discussed above represent the present status of tryptophan production in the world. Nevertheless the international scientific literature is full of contradictions with respect to the economy of the different production processes. Thus an approximate economic calculation from 1977 [108] declares that the production cost of Trp per kilogram would be about 40–80 $ using synthetic methods but only 15–20 $ using a bioprocess (bioconversion from indole), including capital-related charges, too. This contradicts the fact even in 1980 the total tryptophan production derived from synthetic methods, including probably an enzymatic resolution step.

Table 2. Spectrum of tryptophan consumption (kg per year)

Territory	Diets, food products	Parenteral solutions	Drugs	Research	Total
USA	50,600	7,000	0	0.70	57,600
Canada	0	350	0	0.70	350
Common market	1,391	5,000	13,241	1.40	19,632
Japan	695	1,750	139	14	2,599
Latin-America	0	500	139	0	639
Total	52,686	14,600	13,519	16.8	80,820

Though we have seen series of examples of different bioconversion and biosynthesis processes, and theoretically the latter would be more economic, the bioconversion from indole seems to be more feasible. According to Hirose and Shibai [5] "breeding of an L-tryptophan producer (auxotroph, regulatory mutant) with industrial importance has not been successful" and "for some amino acids, e.g. L-tryptophan . . . direct fermentation was very difficult to establish, perhaps owing to the difficulty in avoiding regulatory control".

The cost of production of tryptophan via either a bioconversion or a biosynthetic process would depend on:
1) the cost of the raw materials (medium components and precursor if bioconversion is used);
2) the processing costs of microbial growth (cultivation);
3) the processing cost of the recovery.

The latter two would be probably the same independent of the type of bioprocess; small differences could be attributed to the different cultivation times (there are some processes, both bioconversions and biosyntheses, which need about 50–60 h of cultivation, with 10–12 g l^{-1} product concentration).

The raw material component of the cost is unambigously more advantageous in case of the biosynthesis process, at least because of the precursor. But if a biosynthetic process is still not more economic, the very likely reason for this could be microbiological, i.e. the instability of the multiple mutant microbial strain.

Furthermore, the precursor (indole) price could be relatively low if it was manufactured from less valuable industrial side products (e.g. o-nitro-etil-benzene, o-amino-etil-benzene). Thus a Hungarian estimation predicts 10 $ per kg production cost for indole which means 15–18 $ for one kilogram of tryptophan. With the present high tryptophan prices this seems to be acceptable.

In the future it is very likely that the advanced genetic techniques involved in breeding new and stable producer strains will promise more economical methods for the biosynthetic production of tryptophan. Already, Ajinomoto researchers are using genetic engineering to improve the yields in amino acid production and as an Ajinomoto's scientist says "for some amino acids, costs can be cut significantly" [117]. This technique has cut the cost of producing threonine by 90 % and of lysine by one third.

6 List of Abbreviations

PEP	phosphoenolpyruvate
E4P	erythrose 4-phosphate
DAHP	3-deoxy-D-arabinose-heptulosonic acid 7-phosphate
DHQ	5-dehydroquinic acid
DHS	5-dehydroshikimic acid
SA	shikimic acid
SAP	shikimic acid 5-phosphate
EPSAP	3-enolpyruvylshikimic acid 5-phosphate
CA	chorismic acid
PPA	prephenic acid

AA   anthranilic acid
PRA   N-5'-phosphoribosyl anthranilic acid
CDRP   1-(o-carboxyphenylamino)-1'-deoxyribulose-5'-phosphate
InGP   indole-3-glycerol-phosphate
In   indole
GAP   glyceraldehid 3-phosphate
Trp   tryptophan
Phe   phenylalanine
Tyr   tyrosine
DAHPS   3-deoxy-D-arabinose-heptulosonate-7-phosphate synthetase
            E.C. 4.1.2.15.
DQS   5-dehydroquinate synthetase
DQ   5-dehydroquinase *E.C. 4.2.1.10.*
DSR   dehydroshikimate reductase *E.C. 1.1.1.25.*
SK   shikimate kinase
EPSAPS   3-enolpyruvylshikimate-5-phosphate synthetase
CS   chorismate synthetase
CM   chorismate mutase
AS   anthranilate synthetase *E.C. 4.1.3.27.*
PRT   anthranilate-phosphoribosyl transferase *E.C. 2.4.2.18.*
PRAI   N-(5-phosphoribosyl)-anthranilate isomerase
InGPS   indoleglycerol phosphate synthetase *E.C. 4.1.1.48.*
TS   tryptophan synthetase *E.C. 4.2.1.20.*
P5P   pyridoxal 5'-phosphate

7 References

1. Kinoshita, S.: Adv. Appl. Microbiol. *1*, 201 (1959)
2. Terui, G.: Tryptophan, in Yamada, K. et al.: The Microbial Production of Amino Acids, Kodansha Ltd., Tokyo 1972
3. Esaki, N. et al.: Biotech. Bioeng. *22*, Suppl. *1*. 127 (1980)
4. Hirose, Y., Okada, H.: Microbial Production of Amino Acids, in Microbial Technology, (Peppler, H. J., Perlman, D. ed.) Vol. 1. Academic Press 1979
5. Hirose, Y., Shibai, H.: Biotech. Bioeng. *22*, Suppl. *1*. 111 (1980)
6. Lehninger, A. L.: Biochemistry, Worth Publishers, New York 1972
7. Margolin, P.: Regulation of Tryptophan Synthesis, in Metabolic Pathways, (Vogel, H. T. ed.) Vol. 5. Metabolic Regulation, Academic Press, New York—London 1971
8. Truffa-Bachi, P., Cohen, G. N.: Ann. Rev. Biochem. (1973)
9. Greenberg, D. M.: Carbon Catabolism of Amino Acids, in Greenberg, D. M.: Metabolic Pathways Vol. II. p. 79. Academic Press, New York—London 1961
10. Miles, E. W., Kumagai, H.: J. Biol. Chem. *249*, 2843 (1974)
11. Zaffaroni, P. et al.: Agr. Biol. Chem. *38*, 1335 (1974)
12. Dinelli, D., Morisi, F. (Snam Progetti SPA): German Pat. 2225797
13. Colowick, S. P., Kaplan, N. O.: Methods in Enzymology, Vol. II. Academic Press, New York 1955
14. Aurich, M., Schöpp, G.: Z. Allg. Mikrobiologie *13*, 545 (1973)
15. Matchett, W. H.: J. Biol. Chem. *249*, 4041 (1974)
16. Abbott, B. J.: Adv. Appl. Microbiol. *20*, 232 (1977)
17. Hitoshi, E. et al. (Ajinomoto): US Pat. 3808101
18. Sano, K., Mitsugi, K. (Ajinomoto): German Pat. 2461188

19. Teuscher, G., Teuscher, E.: Acta Biol. Med. German *19*, 211 (1967)
20. Watanabe, T., Snell, E. E.: Proc. Nat. Acad. Sci. USA *69*, 1086 (1972)
21. Yamada, M.: J. Ferment. Technol. *32*, 259 (1974)
22. Lester, G.: J. Bact. *107*, 448 (1971)
23. Shetty, A. S., Gaertner, F. M.: J. Bact. *113*, 1127 (1973)
24. Isono, K., Mino, Y.: J. Japan Grassl. Sci. *16*, 285 (1970)
25. Elander, R. P. et al.: Appl. Microbiol. *16*, 753 (1968)
26. Teuscher, G.: Allg. Mikrobiol. *7*, 393 (1967)
27. Teuscher, G.: Allg. Mikrobiol. *11*, 431 (1971)
28. Doy, C. M.: Rev. Pure Appl. Chem. *18*, 41 (1968)
29. Meuris, P.: Bull. Soc. Chim. Biol. *49*, 1573 (1967)
30. Jensen, R. A. et al.: J. Bacteriol. *94*, 1582 (1967)
31. Gibson, F., Pittard, J.: Bacteriol. Rev. *32*, 465 (1968)
32. Doy, C. M.: Biochem. Biophys. Res. Commun. *26*, 187 (1967)
33. Pittard, J. et al.: J. Bacteriol. *97*, 1242 (1969)
34. Hütter, R., DeMoss, J. A.: J. Bacteriol. *94*, 1896 (1967)
35. Carlton, B. C.: J. Bacteriol. *94*, 660 (1967)
36. Kida, S. et al.: J. Ferment. Technol. *43*, 233 (1965)
37. Maurer, R., Crawford, I. P.: J. Bacteriol. *106*, 331 (1971)
38. Miozzari, G. et al.: J. Bacteriol. *134*, 48 (1978)
39. Baskerville, E. N.: J. Bacteriol. *110*, 146 (1972)
40. Carsiotis, M., Jones, R. F.: J. Bacteriol. *119*, 889 (1974)
41. Lester, G.: J. Bacteriol. *107*, 193 (1971)
42. Miozzari, G. et al.: Arch. Microbiol. *115*, 307 (1977)
43. Schürch, A. et al.: J. Bacteriol. *117*, 131 (1974)
44. Petrov, D. F. et al.: USSR Pat. 310930
45. Petrov, D. F. et al.: USSR Pat. 293845
46. Kaneo, T., Izumi, Y.: Synthetic Production of Amino Acids, Kodansha Ltd., Tokyo 1974
47. Sano, K. et al.: Agr. Biol. Chem. *41*, 819 (1977)
48. Ishii, K., Takanami, K.: Japan Pat. 7886093
49. Okazaki, H.: Agr. Biol. Chem. *32*, 254 (1968)
50. Konosuke, S., Mitsugi, K.: US Pat. 3929573
51. Lim, P. G., Mateles, R. I.: J. Bacteriol. *87*, 1051 (1964)
52. Bang, W. G. et al.: Prep. Eur. Cong. Biotechnol. 1ˢᵗ 1978, 186-9 Dechema
53. Decottignies-Le Marechal, P. et al.: Eur. J. Appl. Microbiol. Biotechnol. *7*, 33 (1979)
54. Sahm, H., Zähner, H.: Arch. Microbiol. *76*, 223 (1971)
55. Hoch, S. O. et al.: J. Bact. *105*, 38 (1971)
56. Prozorov, A. A. et al.: USSR Pat. 408758, 324860
57. Shiio, I. et al.: Agr. Biol. Chem. *37*, 1991 (1973)
58. Thiemann, J. E., Pagani, H.: British Pat. 1298499, USA Pat. 3700558, USSR Pat. 351375, German Pat. 2037763
59. Lim, P. G., Mateles, R. I.: Science *140*, 188 (1963)
60. Bandiera, M. et al.: Experientia *23*, 724 (1967)
61. Shiio, I. et al.: Agr. Biol. Chem. *39*, 627 (1975)
62. Nakayama, K., Hagino, M. (Kyowa Hakko Kogyo Co.): German Pat. 1903336
63. Nakayama, K. et al. (Kyowa Hakko Kogyo Co.): Japan Pat. 50-135284
64. Nakayama, K. et al. (Kyowa Hakko Kogyo Co.): German Pat. 2357100
65. Hagino, M., Nakayama, K.: Agr. Biol. Chem. *39*, 343 (1975)
66. Suzuki, M. et al.: J. Ferment. Technol. *55*, 466 (1977)
67. Kida, S., Matsushiro, A.: J. Ferment. Technol. *43*, 302 (1965)
68. Kida, S., Matsushiro, A.: J. Ferment. Technol. *43*, 307 (1965)
69. Maksimova, E. A. et al.: Khim. Farm. Zh. *11*, 75 (1977)
70. Semenova, L. E. et al.: Khim. Farm. Zh. *12*, 94 (1978)
71. Moran, F., Raboin, D.: Proc. IV. IFS: Ferment. Technol. Today 429 (1972)
72. Shirai, K. et al.: Japan Pat. 77125695
73. Shiio, I., Sugimoto, S.: J. Biochem. (Tokyo) *83*, 879 (1978)
74. Malin, B.: USA Pat. 2999051

75. Malin, B., Westhead, J.: J. Biochem. Microbiol. Techn. Eng. *1*, 49 (1959)
76. Arima, K. et al.: USA Pat. 3801457, French Pat. 2128669
77. Vyskumny Ustav Antibiotik a Biotransformaci: Czechoslovak Pat. PV 7345-78
78. Vyskumny Ustav Antibiotik a Biotransformaci: Czechoslovak Pat. PV 7392-78
79. Sanchez, S., Demain, A. L.: Appl. Environ. Microbiol. *35*, 459 (1978)
80. Aktiebolaget Bofors: French Pat. 2263302
81. Kyowa Hakko Kogyo Co.: French Pat. 2568652
82. Kareva, I. I. et al.: Prikl. Biochim. Mikrobiol. *2*, 625 (1966)
83. Gribiny, P. P.: Poluchenie i primenenie aminokislot, Riga, 1970
84. Terui, G. et al.: J. Ferment. Technol. *40*, 120 (1962)
85. Enatsu, T., Terui, G.: USA Pat. 3296090
86. Enatsu, T. et al.: J. Ferment. Technol. *41*, 500 (1963)
87. Terui, G. et al.: J. Ferment. Technol. *40*, 441 (1962)
88. Paulsson, L. E. et al., (Bofors Ab): Swedish Pat. 1231/1972
89. Sevella, B. et al.: L-Tryptophan Production by Fermentation. Paper Presented at VI. Int. Ferm. Symp. 1980. London, Canada, July 20–25, 1980
90. Pécs, M. et al.: L-Tryptophan Production by Fermentation. Paper presented at 2. Symp. der Sozialistischen Länder über Biotechn. Leipzig 2–5. 12. 1980
91. Sevella, B. et al.: L-Tryptophan. Paper presented at Hung. Ann. Meet. Biochem. Siófok, Hungary 1–3. 10. 1980
92. Pécs, M.: Tryptophan Fermentation. Ph. D. Dissertation, Univ. Of Techn. Sciences, Budapest 1980
93. Aboliny, T. K.: Fermentacija 53 (1974)
94. Ruban, E. L., Ozoliny, R. K.: Aminokisloti mikrobnogo sinteza, Riga 1968
95. Enatsu, T. et al.: J. Ferment. Technol. *41*, 61 (1963)
96. Terui, G., Enatsu, T.: J. Ferment. Technol. *40*, 252 (1962)
97. Niitsu, M. et al.: J. Ferment. Technol. *47*, 194 (1969)
98. Aboliny, T. K. et al.: Prikl. Biohim. Mikrobiol. *6*, 633 (1970)
99. Beker, M. E. et al.: Prikl. Biohim. Mikrobiol. *7*, 103 (1971)
100. Ruban, E. L. et al.: USSR Pat. 299542
101. Ruban, E. L. et al.: USSR Pat. 247315
102. Viesztur, U. E. et al.: USSR Pat. 353955
103. Abolins, T., Helmanis, R.: Upr. Mikrobn. Sintezom. 48 (1977)
104. Abolins, T. et al.: Technologija Mikrobnogo Sinteza 52 (1978)
105. Kida, S. et al.: J. Ferment. Technol. *49*, 390 (1971)
106. Ebihara, Y. et al.: J. Ferment. Technol. *47*, 733 (1969)
107. Kyowa Hakko Kogyo Co.: British Pat. 1222904, German Pat. 1948796
108. Eldib Engineering and Research Inc.: Chemical Market Research Survey on Tryptophan, 1977
109. Eldib Engineering and Research Inc. Summit (N. .): Chem. Mark. Rep. *211*, 22 (1977)
110. Fukuda, D. S. et al.: Appl. Microbiol. *21*, 841 (1971)
111. Partridge, S. M.: Biochem. J. *45*, 459 (1949)
112. Partridge, S. M., Brimley, R. C.: Biochem. J. *44*, 513 (1949)
113. Raminya, L. O. et al.: Fermentacija 133 (1973)
114. Kyowa Hakko Kogyo Co.: French Pat. 2000632
115. Aboliny, T. K. et al.: Fermentacija 124 (1973)
116. Tribe, D. E., Pittard, J.: Appl. Environ. Microbiol. *38*, 181 (1979)
117. Applezweig, N.: Business Week Dec. 14 (1981)

Author Index Volumes 1–26

L.G.Nickell

Plant Growth Regulators

Agricultural Uses

1982. 29 figures. XII, 173 pages
ISBN 3-540-10973-0

Plant growth regulators can be defined as either natural or synthetic compounds that are applied directly to a target plant to alter its life processes or its structure to improve quality, increase yields, or facilitate harvesting.

The author of this book – Vice President of Research and Development at the Vesicol Chemical Corporation, Chicago, Chairman of the Plant Growth Regulator Working Group, and the Treasurer of the American Society of Plant Physiologists – discusses the effects of these regulators on plant functions such as rootinduction, control of flowering, control of sex, and control of maturation and aging. Emphasis is placed on the practical aspects of growth regulators rather than their mode of action. An extensive bibliography is provided which includes reviews by specialists in all areas of plant growth and development, both basic and applied.

Special consideration is given to the current commercial uses of plant growth regulators for a variety of purposes on a number of economically important crops as well as to the direct and indirect financial returns to the grower.

Tables are provided which show the chemical name, common name, trademark identification, code designation, producer companies, as well as plant growth regulatory activity and, where applicable, other biological activities of the regulators discussed in the text.

The book will be of interest to teachers, students and specialists in agronomy horticulture, plant physiology, crop science, nursery science, landscape architecture, pomology, seed technology, providing them with the only modern treatment of the uses of chemicals for the control of plant growth.
(1201 ref.)

Springer-Verlag
Berlin
Heidelberg
New York

G. Habermehl

Venomous Animals and Their Toxins

Based on the translation of the 2nd German Edition "G. Habermehl, Gift-Tiere und ihre Waffen"

1981. 44 figures, 44 tables. IX, 195 pages. ISBN 3-540-10780-0

Venomous Animals and Their Toxins is a unique presentation of our knowledge of the venoms of marine and terrestrial animals who use their "weapons" for hunting and self-defence. The author, President of the European Section of the International Society of Toxicology, has himself done a considerable amount of chemical research in this area, and has been especially active in the elucidations of the structure of salamander venoms.

Although directed primarily to biologists and chemists, this book is also a source of valuable information for tourists who may come into contact with venomous animals. The purpose of the book is not, however, to try and convince the reader that danger from venomous animals lurks around every corner; instead, it shows how to avoid accidents with these animals and how to behave in case one should occur. With its new insights into the modes of action and constituents of different venoms, and into the significance of the venoms for human beings and for clinical application, *Venomous Animals and Their Toxins* will also prove to be of interest to physicians and medical researchers.

C. Fedtke

Biochemistry and Physiology of Herbicide Action

1982. 43 figures, 58 tables. XI, 202 pages. ISBN 3-540-11231-6

Herbicides are part of modern agricultural production systems and contribute therefore significantly to the economy of agricultural products. This book describes the effects of herbicides on the metabolism of higher plants from the viewpoint of the plant physiologist. It addresses both the agriculturist interested in herbicide modes of action as well as the research plant physiologist using herbicides as inhibitors of plant metabolic reactions. These two groups that traditionally have their own background training, publications, congresses etc., regard herbicides from a very different viewpoint: as agricultural and as research tools. – This work provides both groups with the necessary background knowledge in plant physiology and promotes the future use of herbicides as metabolic inhibitors in plant physiological research to the advantage of pesticide and plant sciences. A well-known example is the photosynthesis-inhibiting herbicide Diuron, known to plant physiologists as DCMU, which has become one of the essentials in modern photosynthesis research. Similary, knowledge in other areas of plant metabolism may be advanced by the use of herbicides as specific inhibitors.

Springer-Verlag
Berlin
Heidelberg
New York